Economic Security and High Technology Competition in an Age of Transition

Economic Security and High Technology Competition in an Age of Transition

The Case of the Semiconductor Industry

Eric Marshall Green

Westport, Connecticut
London

Library of Congress Cataloging-in-Publication Data

Green, Eric Marshall.
Economic security and high technology competition in an age of transition : the case of the semiconductor industry / Eric Marshall Green.
p. cm.
Includes bibliographical references and index.
ISBN 0-275-95253-3 (alk. paper)
1. Semiconductor industry—Government policy—United States. 2. Economic security—United States. 3. National security—United States. 4. Competition, International. I. Title.
HD9696.S43U48184 1996
338.4'762138152'0973—dc20 95-7986

British Library Cataloguing in Publication Data is available.

Library of Congress Catalog Card Number: 95-7986
ISBN: 0-275-95253-3

First published in 1996

Praeger Publishers, 88 Post Road West, Westport, CT 06881
An imprint of Greenwood Publishing Group, Inc.

Printed in the United States of America

The paper used in this book complies with the Permanent Paper Standard issued by the National Information Standards Organization (Z39.48-1984).

10 9 8 7 6 5 4 3 2 1

To Dana

Contents

Preface

This work has been motivated by an awareness of the ever-growing importance of technology on productivity and power in the information age. An underlying premise is that in an era of diminished military confrontation, economic power and technology are certain to acquire enhanced importance in national security considerations and that this is bound to promote closer coordination between government and private industry. With an economic focus and a public policy perspective, this book is designed to clarify the debate on high technology industrial policy, and to address the policy question of whether and how government should respond to slippage in a strategic industry. The text examines the relationships among national security, economic competition, and technology with a broader view toward deriving an appropriate role for government in high technology competition. It is beyond the scope of this study to consider in detail the many high technology sectors, such as aerospace, medicine, composite materials, and biogenetics, that make important contributions to economic and social welfare. The objective of this book is to employ the case study as a test for other high technology sectors that share common challenges embodied in the economics of R&D, economies of scale, compressed obsolescence, and the imperative to service the widest range of markets to recover large front-end investments in research and development.

The semiconductor industry was selected as a case study for several reasons. First, semiconductors provide the building blocks for dozens of manufacturing technologies, drive innovation and product development in electronic commercial technologies, and are essential components in advanced electronic military systems. Second, for many Americans the semiconductor industry symbolizes high technology. The industry first developed in America, it remained dominated by American firms until the early 1980's, and it confirms that the competitive challenge to American industry is not confined to low-wage or low-tech industries. Third, the industry provides a framework to examine the effect that policies of foreign governments have had on competition in the industry, and in light of those policies, the wisdom of industrial policy alternatives for the American industry. It symbolizes the challenge of remaining competitive in a high technology industry where commercial competitors have the benefit (or liability) of government support.

There are many people to whom I owe a deep sense of gratitude for making this work possible. To the Eisenhower Institute of World Affairs who provided me a fellowship to pursue this work to its culmination I am deeply grateful. Dr. Robert Pfaltzgraff of the Fletcher School of Law and Diplomacy and Institute for Foreign Policy Analysis provided valuable expertise, time, and energy. My father-in-law, Dr. Delwin Roy, deserves special thanks for expending an enormous amount of time in helping me re-work successive drafts of this book. His interest and effort was greatly appreciated and beyond any call of duty. Mr. William Gsand, Executive Vice-President of Hitachi America, Mr. Richard Sanquini, Senior Vice-President of Business Development at National Semiconductor, and Mr. Robert White, Under Secretary of Technology in the Bush Administration, all graciously offered time and insight to help conjoin the world of theory to the practical exigencies of private-industry. I am also grateful to the research division of Hitachi America for their assistance in collecting data and industry publications not easily accessible to the general public. A special thanks must be extended to my mother and father for their encouragement and support. They infused me with an enthusiasm for the world of ideas from an early age, without which this book would not have been possible. As a widely published academic, my father also provided valuable assistance and ideas throughout the writing of this manuscript.

My deepest sense of gratitude is reserved for my loving and beautiful wife Dana. She alone assumed the responsibility of providing an income during our first years of marriage, and her unwavering support and faith in me is a continued source of inspiration. In the hope that this work is worthy, I dedicate this book to her.

Eric Green
Federal Reserve Bank of New York
1995

Economic Security and High Technology Competition in an Age of Transition

Introduction

The end of the Cold War marks the beginning of a new world order that makes different demands on national power. Military power remains a vital element in the exercise of security and influence in the wider world, but its value has been substantially diminished by the elimination of the Soviet Union. In an era of reduced military confrontation, national power rooted in domestic economic strength has acquired greater significance in national security considerations. In this context, the advanced technology enterprise has assumed a central importance given its role in economic growth, productivity, and its "dual use" characteristics. At the same time however, concern that the United States has suffered a sustained decline in high technology product markets has focused much attention on America's ability to meet the demands of this different security framework. Although a convergence in technological sophistication among the advanced economies was to be expected, there is a perception that American performance in this area of presumed hegemony has suffered because of industrial policies implemented by foreign governments. Government promotion of high technology industries has long been accepted practice in Europe and Japan, but the United States has questioned the merit of government intervention to promote selected industries, especially where there was no specific military priority. However, given the complexion of economic competition and security in a single-superpower world, the United States has begun to rethink its traditional reverence for laissez-faire economics.

With the reduction in military threats, competitiveness in nonmilitary matters is increasingly important if the United States is to safeguard its leadership position in a new world order. Chapter 1 examines economic power as it relates to national security since 1945. It provides the reader with an assessment of the changing conditions affecting national power and the growing importance of technological and economic factors in the calculus of national security. The analysis is guided by the premise that every government is involved, to a greater or lesser extent, in regulating economic activity, but that the American government has generally,[1] as a matter of ideological preference, avoided sector-specific promotion of economic activity unless it has been related to national security. This chapter questions whether the slippage in economic power in high technology industries should be considered a threat to national security. Accordingly, the chapter examines whether government should elevate economic issues in the calculus of national security to justify government intervention in economic sectors of particular relevance to economic as well as military security.

It may seem self-evident that a competitive position in high technology is desirable because, for an advanced industrial economy, it is the foundation on which future economic prosperity is based. However, a more complete accounting of the impact that certain high technology industries, have on the rest of the economy and on the military establishment is fundamental to any consideration of government promotion. Chapter 2 examines the role of technology as it relates to economic power and military strength. The "strategic" industry concept is examined, high technology is defined, and an analysis of the putative benefits and rents of high technology industries is explored to determine whether this form of economic activity warrants special attention. Building upon the security analysis in the preceding chapter and the strategic-industry notion presented here, the relation between the military and the high technology establishment is considered. The chapter presents an alternative technological-development trajectory, and the ramifications of reliance on foreign state-of-the-art components rounds out the analysis on technology and economic security.

The trend toward internationalization of research and production activities raises questions about the ability of national policy tools to secure international advantage in particular industries. Chapter 3 examines the relevance of a national agenda given the "globalization" of the firm, and it provides the reader with an understanding of competing perspectives in the industrial-policy debate. The rationale for an industrial policy must extend beyond a determination that high technology industries are somehow critical to national security. Such industries may be important, but government intervention may not be the appropriate mechanism to enhance competitiveness. Government promotion may be theoretically possible, but it may not be practical or desirable or feasible. A critical survey of literature on industrial policy and national competitiveness is used to acquaint the reader with the issues that foster disagreement over the necessity and desirability of such policies and the potential pitfalls and liabilities to government promotion. Political and economic justifications for and against government intervention precede conflicting interpretations of past government policy in high definition television (HDTV) and the former Reconstruction Finance Corporation. This survey lays the groundwork for subsequent analysis of the semiconductor industry.

Advances in technology, communication, and transportation are collapsing national markets into a single global medium, and industrial policy cannot, therefore, be conceived without reference to related trade issues. Export subsidies, tariffs, quotas, and other instruments of state-led development efforts are important features of government intervention. Building upon the analysis in the preceding chapter, Chapter 4 examines strategic trade policy and other currents in international economic theory as they relate to national government strategies in international trade competition. This material balances the issues debated on the national level in Chapter 3 with an international economic perspective. Similar problems of practical implementation are examined with special emphasis on profit-shifting models and infant-industry protection.

Chapter 5 commences the semiconductor case study in light of the foregoing chapters. It examines the type and extent of semiconductor promotion in Japan and Europe, with particular emphasis on trade barriers, government subsidy, and the influence of industrial structure on international competitiveness. A short history of global semiconductor competition and a brief overview of country-specific industrial policies in Europe since 1974 precede an analysis of the European industry from a pan-European perspective. The evolution of Japanese industrial policy for semiconductors since 1971 is then analyzed. The European and Japanese successes and failures are subsequently examined in a comparative context focusing on the relative scope and impact of government policies, their evolution, and where and why government policy has yielded the greatest and least results.

Building upon the analysis in Chapter 5, Chapter 6 examines the competitive position of the American industry in relation to its principal rival, the Japanese semiconductor industry. It explores the formation and rationale of America's industrial policy forays in the industry with the technology consortium SEMATECH, and the Semiconductor Trade Agreements. The chapter examines the extent to which the decline of the American industry in the 1980's stemmed from foreign industrial targeting or from other factors that have a significant impact on semiconductor competition. More detail on forces shaping competitive advantage in the semiconductor trade and technology is provided. A comparative analysis focuses on respective trade regimes, breadth of serviceable markets, industrial structure, demand factors, technology development, and capital formation. The reader gains an understanding of the relative strength and structure of the American and Japanese industry, the nature, scope, and efficacy of American industrial-policy initiatives, and their role in the rising fortunes of American semiconductor producers in the 1990's.

The semiconductor industry represents a test case for other high technology industries that share similar characteristics of short product cycles, large R&D requirements, economies of scale, and accelerating development costs. The varied strategies and experiences of the United States, Europe, and Japan illustrate that the overall utility of government intervention is context dependent upon related factors influencing competitive position. In light of the foregoing analysis, concluding Chapter 7 examines whether the United States government is best advised to follow a laissez-faire strategy, or whether there is evidence to suggest that specific forms of government intervention would be more likely to ensure successful competition in high technology markets. The study opened the policy question of how the American government should respond to competitive slippage in a strategic industry. This chapter draws together the principal elements articulated in preceding pages and offers a policy prescription in light of the lessons learned from the semiconductor case study and changing national security imperatives.

The challenge of any study on industrial policy hinges on the weight of empirical evidence, since there is no literature that establishes a direct causal connection between industrial policy and success or failure in high technology. It is difficult to disaggregate the effects of government policies from other macro-level or industry-specific factors that relate to competitive performance. Part of this problem stems from the inherent difficulty in measuring the effect and extent of externalities. Comparing the levels of competitiveness in an industry is a complicated task because trade protection, subsidies, and different accounting conventions can cause many presumed measures of competitive advantage to be misleading. For our purposes, market share and technological innovation are used as yardsticks of competitiveness because both are essential to long-term success in high technology competition, and it is assumed that these positions are maximized. (Technological innovation may also refer to semiconductor manufacturing systems that provide an advantage in lower technology semiconductor components.) Technological innovation is a constant necessity owing to short product life cycles. Also, since economies of scale encourage firms to service the widest breadth of markets, it is assumed that market-share can indicate whether the national industry is gaining or losing in international competition. Standard practice is to base market-share figures on dollar-denominated revenues for each country divided by the total dollar-denominated industry revenues, and this is the technique used for market share figures throughout the text.[2]

This work adopts a structural perspective toward national competitiveness.[3] It differs, in some measure, with the orientation that any theory of national competitiveness must begin at the firm level because the principal economic issue for any nation is productivity, and the source and rate of productivity growth is determined by firms. A company-level approach to competitiveness certainly has merit. It is useful to explain how firms differentiate themselves in market competition against other firms. However, this orientation fails to account for the relative performance of national industries in international competition—why some strategies work in some countries and not in others—and it does not explain the role assumed by specific national strategies in international competition. National technological capabilities have been and continue to be an important variable in national competitiveness. The importance of technological change in economic growth has been established in a stream of economic literature;[4] international differences in innovative capabilities are important in explaining differences in country growth rates. The structural perspective adopted in this study holds that the fate of high technology industries may be less dependent upon company-level actions than it is upon the competitive structure in which the firm operates.

NOTES

1. The term generally is used because, as Chapter 3 indicates, the government has been involved in sectoral promotion in areas that are not of obvious relevance to national security.

2. Market share calculations are therefore sensitive to exchange rate fluctuations. A high dollar will underestimate the share held by foreign firms and overestimate the share held by American firms, and vice versa. During the first half of the 1980's when the dollar appreciated against all major currencies the decline in American market share was more pronounced than prevailing figures. The opposite is true during the era of a low dollar after the mid-1980's.

3. For a definition of national competitiveness see Competitiveness: A Shrinking Asset in Chapter 1.

4. Nathan Rosenberg and David Mowery, *Technology and the Pursuit of Economic Growth* (Cambridge: Cambridge University Press, 1990); Edward Denison, *The Sources of Economic Growth* (Washington, D.C.: Committee for Economic Development, 1962); Jan Fagerberg, "Why Growth Rates Differ," in *Technical Change and Economic Theory*, Giovanni Dosi and others, eds. (New York: Francis Pinter, 1988); Raymond Vernon, ed., *The Technology Factor in International Trade* (New York: Columbia University Press, 1970).

1

Economic Power and National Security

THE COLD WAR: DEFINITION AND PRACTICE OF NATIONAL SECURITY

One may consider any national security policy as one in which government policy is calculated to create national or international political conditions favorable to the protection and extension of vital national values against existing or potential adversaries. Fundamental to the implementation of a security policy is the assumption that the primary unit in international politics is the nation state and that each state determines the type of political and economic system appropriate to the preservation and enhancement of national values. What constitutes core values can differ in time and between states. The emphasis and urgency afforded to either political, economic, social, or military values at any one time reflects the relative strength of the state in the international system and its need to preserve or secure advantage. Although values can be difficult to define and articulate, it is clear that a value common to all nations is a desire to survive, both in a physical sense and as a way of life. As leader of the free world during the Cold War, the United States was compelled by necessity to assume a large military burden in the interests of free world security. Preserving national security in that context was dependent on a large military capability.

National survival values may assume various forms. Survival via objectives of political independence and territorial integrity tend to be more defensive, while variations including territorial and ideological imperialism are more aggressive and extension oriented. What is clear is that values tend to be somewhat ambiguous, vague, and, therefore, rarely suitable as guidelines for the daily conduct of national security policy. Intermediate objectives, termed "interests," provide a link between ultimate values and the techniques of national security operations. Interests such as the containment of Soviet communism during the Cold War are not ultimate objectives but a means to secure the values articulated by the government. Interests are subject to more immediate change. Like values they can be confusing and difficult to articulate; over time they can become institutionalized and assume a life of their own apart from the values they were intended to support. Interests that are not tailored to reflect a changing security environment may be indicative of institutional rigidity. Such a condition may be difficult to overcome.

The traditional American approach to national security was influenced by natural geographic advantages and the popular dislike for large standing armies. Despite occasional lapses to the contrary the United States pursued isolationist policies and avoided entangling alliances that might, in the American view, drag the country into war. This orientation was dramatically altered by World War II, however. Nuclear weapons, communist competition, and America's unrivaled status as industrial superpower undermined the traditional sense of continental security and induced the United States to acknowledge that the preservation of national security required a new global focus and constant military preparedness.

The United States redefined its national security interests as the principal security imperative in the postwar world became the containment of Soviet communism. Military power generated by a vital and integrated economic system would provide the muscle for containment. The Truman Doctrine institutionalized the departure from the American tradition of minimal peacetime involvement by providing for direct American intervention in the internal affairs of other states whenever they were threatened by loss of independence through communist aggression. It was the foundation for containment. The Marshall Plan (1947) represented America's economic commitment in the postwar system. By facilitating the rehabilitation of West European economies, the United States secured markets for its goods, effectively rendered communism an unattractive alternative, and provided its Western Allies the long-term economic resources to support an expanded military mission.

The creation of NATO in 1949 formalized the shared commitment of the Western Allies. The United States later extended its policy of containment to regional networks, including ANZUS, SEATO, and CENTO, thereby pledging U.S. economic and military support to governments that shared America's commitment to contain communism. These international security arrangements and adherence to the collective-security concept of the United Nations meant that the United States was thoroughly enmeshed in a global security network. The prevailing view was that the very survival of capitalism and American democracy depended on containing the communist threat.

At first, the U.S. atomic monopoly enabled the United States to forgo any immediate systemic and theoretical analysis associated with the demands of containment. However, the Soviet acquisition of nuclear weapons, the ascendancy of communism in China, and the outbreak of the Korean War necessitated a review of national security policies. The result was National Security Council document 68 (NSC-68), the first comprehensive analysis and synthesis of a U.S. national security strategy after 1945. The document detailed the U.S. defense interests against communist aggression and called for a marked increase in defense expenditure for conventional forces. Containment required a sustained military capability, one reflected by military preparedness and forces in being rather than a vague sense of mobilization potential.[1] It was in the spirit of this commitment that America's enormous military-industrial complex was conceived.

The economic burden imposed by the postwar security framework was not an obvious concern. The United States emerged from World War II as the only industrial superpower. While its natural economic competitors suffered from the ravages of war, the United States entered the new era with its industrial establishment intact and running in high gear. The GDP of the United States represented almost half of total world output. America was unchallenged in its high technology leadership, its capital and human resources, its efficiency and productivity, its natural resources, and its capacity to engage in world trade. American industry had won the war and it would anchor the new international economic system. Given its immense economic power, the United States could afford to tolerate deviations from free trade and currency convertibility in the interest of economic and political rehabilitation of its allies. It was an era of unique American prosperity. Competitiveness was not cultivated. It was assumed.[2]

An important component of the postwar security system was the establishment of an open and liberal economic order. The United States argued against imperial preference, bilateralism, predatory trade practices, and the "us versus them" mentality that had colored international economic relations prior to the war. Economic interdependence would, it was argued, give all the industrial economies a stake in the system and thereby discourage the resort to armed conflict. The new system engineered at Bretton Woods in 1944 was founded on multilateral free trade and institutionalized in the International Monetary Fund (IMF), the International Bank for Reconstruction and Development (World Bank), and the General Agreement on Tariffs and Trade (GATT). As the world's preponderant economic power, the United States provided the liquidity in the international financial system and the political stability to liberalize the old order.

The industrial economies of the world flourished between 1950 and 1973. The Bretton Woods system encouraged international trade and investment and nourished a liberal trading philosophy such that world trade expanded at a faster rate than world output.[3] This growth in world trade, reflecting the technological revolution in communication and transportation, as well as increasing capital mobility, made the international economic system decidedly more interdependent.[4] However, this also made domestic economies more sensitive to economic developments overseas. The Bretton Woods mechanism of fixed exchange rates was abandoned in 1973 for flexible exchange rates in large measure to isolate domestic economies from international economic disruptions. Although the failure to maintain the dollar peg signaled the end to the Bretton Woods system, the liberal orthodoxy of free trade and investment remained the cornerstone of the international economic order, at least as official policy.

During those years Germany and Japan made spectacular strides in rejoining the ranks of the major economic powers, and by the 1970's they had become the principal commercial rivals of the United States. Over the decades of the Cold War, America's allies, particularly Japan and the northern countries of the European Community, restored economic stability; developed highly sophisticated national infrastructures; invested in education, research, and development; and achieved a high degree of competitiveness in industrial production. The subsequent convergence among the most advanced industrial nations in per capita income and in output per-man hour suggested that the unparalleled international economic advantage enjoyed by the United States at the end of World War II had begun to evaporate. The maturity of America's commercial competitors and the manifestation of vigorous export-driven economies in the newly industrialized countries along the Pacific Rim, together with the attenuation of industries considered important to national security, have fostered the growing public debate over economic security.

Convergence may be explained in part by the artificially high economic position of the United States compared to its war-ravaged competitors after World War II. A more realistic alignment was to be expected as those economies were rebuilt and retooled for economic competition. The enormous gap in economic power was bound to narrow. However, many analysts regard this convergence as evidence that American industry is losing its competitive edge. There has been a proliferation of writings in recent years on what many perceive to be the decline of the economic prosperity and international power of the United States.[5] Most of these works claim that the United States, following the example of nineteenth century Britain, has lapsed into a terminal period of economic lethargy. Some attribute this to the burden of defense spending,[6] others to the rise in other nations of new and better ways of organizing economic activity.[7] Yet while slippage in the American economic position is widely recognized, the leading position of the United States has not changed fundamentally. If the United States has declined compared to the artificial high occasioned by World War II, then it is logical to assume that convergence may be less a manifestation of decline than a reflection of America's returning to a more realistic long-term equilibrium with its principal economic rivals.

COMPETITIVENESS: A SHRINKING ASSET?

There has been an enormous amount of attention in the last several years devoted to the concept of national competitiveness. This preoccupation stems from an attempt to better understand the changing dynamics of global competition between rival economies in an era characterized by joint ventures and cross-border business alliances. As such, it is a concept that challenges any simple attempt to define it. Economists and policy makers have many measurements of economic activity to monitor the vital signs of economic performance. A sample would include gross domestic product (GDP), real wage growth, and productivity. Although useful, these indicators do not fully reflect the contemporary debate on national competitiveness because they are inadequate guidelines for analysis on the micro level. The premise that economic problems confronting a nation are similar to those that challenge the firm is debatable, but it does touch upon an important element of competition. A firm that is competitive is one that is able to increase earnings by expanding sales and/or profit margins in the market, and a firm that is uncompetitive will not be able to maintain its market position and will eventually be forced out of business. A nation, on the other hand, cannot go bankrupt and there is no well-defined bottom line as there is for a firm. Nevertheless, national and company-level concepts of competitiveness, while analytically distinct, are functionally intertwined. The ability of a nation to provide for an expanded standard of living will almost certainly depend upon competitive firms generating the productivity levels needed to support higher wages. That competitiveness in turn depends upon the quality and quantity of physical, human, and capital resources; the manner in which those resources are managed; the supporting infrastructure of the economy; and government policy.

The Young Commission's report, *Global Competition: The New Reality*, examined national competitiveness in light of America's performance in world markets. The report defined competitiveness as the "degree to which a nation can, under fair market conditions, produce goods and services that meet the test of international markets while simultaneously maintaining and expanding the real incomes of its citizens."[8] This competitiveness forms the basis for a nation's standard of living and the capacity to meet international obligations. Yet national competitiveness as measured by performance on international markets is potentially misleading. In both theory and practice, a positive trade performance may actually reflect a systemic weakness rather than a productive strength.[9] Performance in international markets is revealing when comparing firms or industries, but it offers only one glimpse of a nation's ability to compete in the international arena.

Prioritizing indicators of national competitiveness remains a subject of debate. Yet any analysis of national competitiveness should be textured to accommodate the performance of domestic productivity, because it is the capacity to produce more goods and services with constant or diminishing input values that will ultimately determine one's ability to compete globally. Domestic productivity, as measured by output per average worker, is determined by a complex array of factors, many of which are beyond the purview of government policy. However, it is directly affected by the composition of national resources such as education, capital formation, economic infrastructure, and even crime. These factors not only affect our ability to produce goods efficiently but influence what goods are produced. The absolute level of productivity among competitor nations carries important implications. On the one hand, higher productivity and expanding real incomes in a foreign nation are beneficial insofar as they increase the nation's ability to purchase our goods. On the other hand, should their ability to capture high value-added industries negatively affect America's ability to participate in those same industries, then international trade will produce winners and losers.

While there is agreement that the United States has suffered slower productivity growth than its competitors, there is less agreement over the specific causes of this decline. It should be noted, however, that the role of the firm cannot be overlooked. Organizational and attitudinal deficiencies are very much responsible for the decline in American competitiveness and productivity growth. The annual *World Competitiveness Report*, published by the World Economic Forum, a Swiss organization, ranks the competitiveness of firms in twenty-three different countries on a variety of issues, and the United States has consistently ran 12th to last place (23rd) while Japan and Germany consistently have achieved first or second ranking.[10] The slow response of American business to the realities of global competition[11] and a general disdain for planning beyond the short term have compounded the productivity dilemma. In addition, changes in technology and manufacturing techniques are increasingly pioneered by foreign competitors.[12] This declining competitive position in international markets has not been lost on American business managers. Over the past several years corporate America has undergone significant restructuring to reduce waste and improve efficiency, the results of which are captured in the 1994 *World Competitiveness Report*, which places American firms first among those nations surveyed. Private initiatives to reorganize the corporation should be regarded as an important first step in improving overall national competitiveness. It should be noted, however, that competitiveness rankings do not indicate that a nation is competitive in those high-income producing industries, low technology or high technology, that will determine a country's standard of living.

Compelling arguments have been offered by leading economists that process innovation and strategy are formulated at the level of the firm and that America's competitiveness begin there.[13] Competitiveness on the company-level depends, among other things, upon advanced production techniques, the creation of new products and/or services, and successful commercialization of those same products and services. Innovation must be complemented with a production process that ensures that the innovator is the most cost-effective producer. This is an operational axiom not captured in competitiveness rankings, and there is some evidence that many American companies have not absorbed the advanced production technologies required to be competitive.[14] In this regard, government policy cannot provide a panacea for competitive advantage. Government can prime the innovative pump with subsidies targeted at specific industries or technologies and influence the competitive framework in which the firm operates, but ultimately the utility of such policies will be limited unless appropriately capitalized on by firms in the industry that the policy was conceived to assist. Since the private sector alone can develop, produce, and market new technologies in products and processes, company-level explanations indicate that at least part of the competitiveness problem lies inside the firm.

Since the end of World War II the United States government has extensively promoted basic research across a broad scattering of disciplines in the science and humanities, in part to gain the upper hand in the Cold War, which placed a premium on technological sophistication, and in part because of the belief that such advances would improve the nation's ability to meet the economic and social challenges of the modern world. This view guided America's extensive commitment to basic research after World War II. By adding to the system of innovation in which American firms are linked, the fruits of basic research, commercialized by industry, would raise overall economic and social welfare. The U.S. government has promoted the world's most extensive system of basic research. It provides immense support to the nation's university system, and it is unrivaled in matters of science, as indicated by the award of Nobel prizes. However, translating that technology into commercial products has often been achieved best by foreign competitors. Until recently, the principal policy of Japan's Ministry of International Trade and Industry (MITI) was not to facilitate the extension of technological horizons but to complete and refine technologies for which the basic research had been undertaken elsewhere. MITI's research priorities stressed the development of generic production processes that enabled firms to convert emerging technologies into low-cost, high-volume output. As one observer commented, "while the Americans collected Nobel laureates the Japanese collected markets."[15]

The American position in world trade has gone from one of perennial surplus to one of perennial deficit. The merchandise deficit is the focus of our attention, as the overall trade balance is colored by issues such as investment flows that do not intrinsically reflect a nation's industrial competitiveness or its ability to raise the standard of living.[16] Between 1945 and 1971 the United States did not register a single merchandise trade deficit. Since that time, with the exception of 1973 and 1975, the United States has experienced annual trade deficits. In the 1980's and 1990's those deficits became enormous. The American current account deficit grew to $144 billion in 1987, decreased to $92 billion in 1990, and accelerated over $120 billion in 1993 and 1994. By contrast, Japan had a peak surplus in 1987 of $87 billion. It dropped to $36 billion in 1990, but through 1993 and 1994 it was running at an annual rate in excess of $100 billion.[17]

An examination of the trade deficit is incomplete without accounting for the value of currency. Although an exact accounting is not possible, it is widely believed that an extraordinarily overvalued dollar in the early 1980's promoted imports and priced American goods out of many world markets. Efforts to reduce the overvalued dollar culminated in the Plaza Accord of 1985, in which the G-7 nations pledged to support a dollar devaluation. Subsequent devaluation of the American dollar by a factor of two against most major currencies did result in better terms of trade. The United States had large gains in resource-intensive manufactured products (steel and paper), traditional manufactured products (textiles, apparel, furniture), and scale-intensive manufactured products (chemicals). However, many producers did not regain lost market share and the position of some advanced technology sectors, notably electronic components, semiconductors, and even computers, continued to deteriorate.[18] Despite a sharp appreciation in the yen and depreciation of the dollar, the Japanese share of world high technology exports continued to rise and the American share continued to fall. In some measure this may be explained by *pass-through,* as many Japanese industries passed through the affects of a revalued yen by lowering the price of its goods to retain existing market share.[19] In addition, Japanese and European producers reduced production costs owing to artificially cheaper oil. With a higher dependence on foreign oil, European and Japanese industries benefited because as the dollar fell, so did the price they had to pay for oil, an international commodity denominated in American dollars. Efficiency gains from cheaper oil may provide some insight into the competitive position of resource-intensive industries, but it does little to explain the persistent decline in advanced technology sectors such as electronics throughout the 1980's, an industry in which energy is a minor factor input. Moreover, it was in resource-intensive industries that American producers regained lost market share after 1985. World market-share figures are calculated by using dollar-denominated revenues for each country divided by total dollar-denominated industry revenues. Given the high value of the dollar, this implies that the dwindling market share for electronics and semiconductors in the early 1980's was even greater than the observed decline. It also suggests that the realignment of the dollar since 1985 has tended to decrease the actual American market share. Whether or not this discrepancy is partial or fully offset by larger exports by the country with the weaker currency is difficult if not impossible to determine. On balance, however, the persistence of large American trade deficits through the 1990's in light of a cheap dollar suggests that American industry may simply be less competitive in international markets.

Sustaining a trade deficit requires a huge capital inflow from the rest of the world. The United States began the 1980's as the largest creditor nation, but has since become the largest debtor nation and faces the prospect of paying 4 to 5% of its GDP in debt servicing to domestic as well as foreign creditors. Large-scale capital borrowing is not necessarily bad provided the capital is invested in the productive resources of the country, thereby facilitating the servicing of accumulated debt. However, perennial trade deficits are an indication that the United States is consuming at a level not commensurate with its economic performance. At the same time, a nation's current account will, by definition, balance out. This means that those countries with surplus dollars must find a way to spend those dollars in the United States, and during the 1980's, many widely publicized acquisitions by Japanese corporations of American companies, real estate, and government securities fostered public paranoia over a new brand of imperialism. In addition, the U.S. government has borrowed to finance government consumption (i.e., military buildup and entitlement programs). This external debt constrains American fiscal and monetary policy because political and economic considerations prevent the government from choosing a course of action that would either add to the national debt or jeopardize the capital infusion on which the trade deficit (and also the budget deficit) is financed. In fact, for reasons pertaining more to the recession at home, since 1991 Japan has repatriated profits from foreign operations and is now a net importer of foreign capital. In this case, the current account is being balanced with the United States through the appreciation of the Japanese yen. Despite this trend, however, the American trade deficit with Japan has not been reduced, a development that is inconsistent with the basic mechanisms of international trade—that is, one would have expected Japanese imports into the United States to be priced out of the market. When countries repatriate their surplus capital, it is still a problem for the United States because exposure to a falling currency value against the dollar will affect the level of investment in U.S. government securities, as they will eventually require a higher premium to offset losses to the portfolio when translated into their home currency.

The availability of capital resources is an important factor of competitiveness on which private firms depend. Recent studies indicate that the cost of capital in the United States remains higher than it is in Japan.[20] A higher cost of capital means that companies must generate more wealth before breaking even in any investment.[21] Higher capital costs increase risks in financing decisions and thereby affect the aggregate level of investment as well as the planning horizons of a firm. Because investment in productive assets is diminished, American companies may suffer a competitive disadvantage. The globalization of capital markets is steadily making real capital costs more equal among nations, but national differences remain significant and will persist.

Traditional macroeconomic indicators such as GDP growth, real wages, and domestic productivity continue to be important measurements of economic power, and they provide a mixed record for American competitiveness. Since 1973 real wages have declined both relatively and absolutely.[22] Table 1.1 presents the gross domestic product growth of countries belonging to the Organization for Economic Cooperation and Development (OECD) from 1961 to 1993. Between 1971 and 1980 the United States had the fifth lowest growth rate but outperformed West Germany. Between 1980 and 1990 the average growth rate for the United States (2.71%) had been below that in Japan (4.34%) but higher than Germany (2.17%) and every other European economy. Real GDP indicators do not provide any evidence of declining American competitiveness.

Productivity is the central factor determining the ability of any society to generate a high standard of living. The United States remains the most productive nation in the world, thanks to its skilled resources and technological diversity across a broad spectrum of industries.[23] Figure 1.1 indicates that

Table 1.1
Real Gross Domestic Product Growth, 1961-1993

Country	1961-70	1971-80	1981-85	1986	1987	1988	1989	1990	1991	1992	1993
United States	3.8	2.8	2.6	2.8	3.4	4.4	2.5	1.0	-1.1	2.4	2.8
Australia	5.2	3.4	3.1	1.9	4.3	3.9	4.7	1.6	-1.0	2.1	2.5
Canada	5.2	4.5	2.8	2.9	4.2	5.6	3.0	.1	-1.7	.07	2.5
Japan	10.5	4.7	4.0	2.7	4.6	5.7	4.8	5.6	3.8	1.3	-.5
Sweden	4.6	2.0	1.8	2.3	2.9	2.3	2.1	.3	-1.1	-2.0	-2.7
Switzerland	4.7	1.2	1.4	2.9	2.0	2.9	3.5	2.6	.1	0	-.8
Belgium	4.9	3.2	.8	1.5	1.7	1.1	1.2	-15.0	1.8	.8	-1.2
France	5.6	3.6	1.4	2.5	2.2	3.8	3.6	2.8	.6	1.5	-.9
Germany	4.5	2.7	1.2	2.2	1.5	3.6	3.8	4.6	4.4	1.4	-2.1
Italy	5.7	3.8	1.4	2.7	2.9	4.2	3.2	2.0	1.2	.9	-.1
Netherlands	5.1	2.9	1.0	2.0	.8	2.7	4.0	3.1	2.0	1.4	-.2
Spain	7.3	3.5	1.4	3.3	5.6	5.2	4.9	3.7	2.1	.7	-1.0
United Kingdom	2.9	1.8	2.1	4.5	4.2	4.0	1.7	.8	-2.4	-.4	2.0

Source: Main Economic Indicators, OECD, April 1994.

Figure 1.1
Labor Productivity as Percentage of U.S. Labor Productivity, * 1950-1991

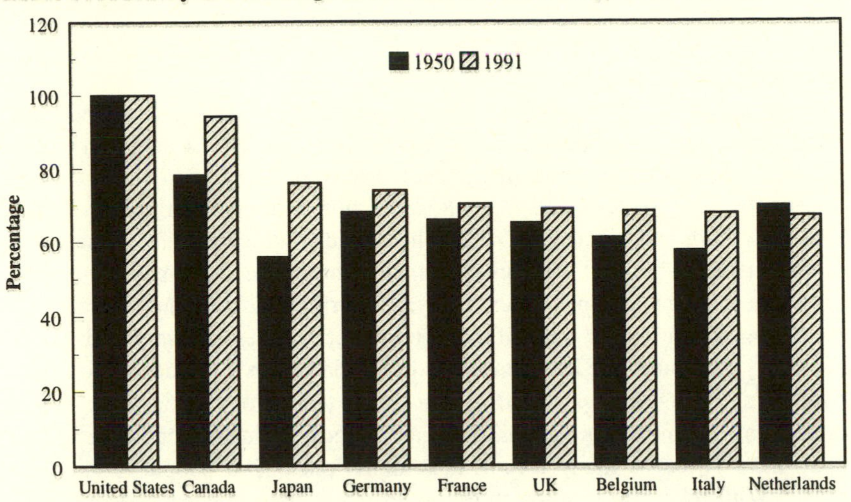

*U.S. Productivity is shown as 100% in each year because the United States remains the productivity leader.

Figure 1.2
Increase in Labor Productivity, 1950-1991

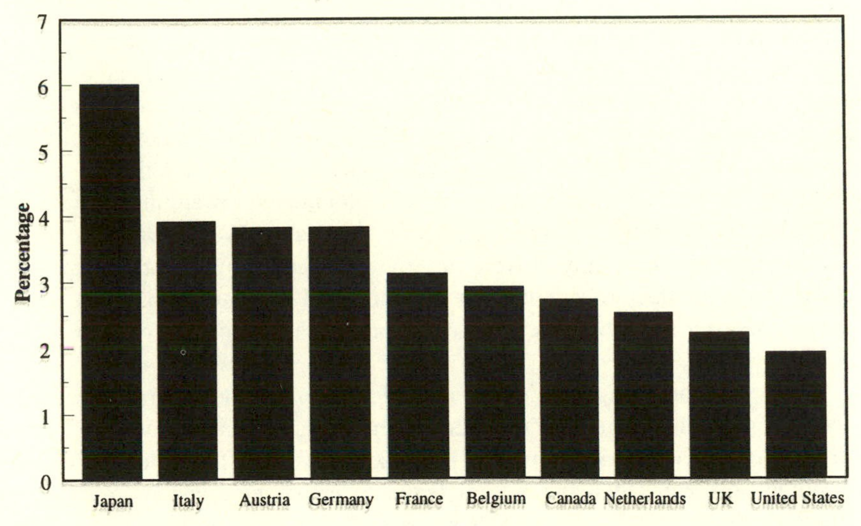

Source: U.S. Department of Labor

American labor productivity remains the highest in the world. However, as Figure 1.2 indicates, the rate of productivity growth has been greater among our commercial rivals. From 1980 to 1990 American productivity grew at the annual rate of 1.2%; in Japan it grew at 3.4%, in France 1.9%, in the United Kingdom 2.8%, and in West Germany 1.4%.[24]

It is tempting to believe that the convergence in productivity is a function of falling American standards in education and economic infrastructure. The U.S. has the highest illiteracy rate in the industrialized world. American students persistently perform less well than those of our commercial rivals in critical areas of math and science.[25] Current trends indicate that America is compromising its ability to capture the more technical and high-paying jobs that contribute most to national productivity. Nevertheless, the United States still retains the largest pool of scientific and engineering talent in the world. Indeed a study by the European Roundtable of Industrialists in 1988 concluded that a problem affecting European efforts in high technology industries was an insufficient technical labor force. The same study indicated that America and Japan have twice as many technical students per 1,000 people in the labor force.[26] With regard to infrastructure, government investment has decreased in the United States and remains below the level of investment in Europe and Japan. One study suggests that perhaps 60% of the decline in productivity growth can be attributed to reduced levels of public investment,[27] and another by the same author indicates a historical trend where an investment increase of 1% in infrastructure translates into a .5% increase in private productivity growth.[28] Education and infrastructure are two obvious areas where government policy can increase productivity over the long term, but this must be balanced with costs associated with higher taxes or larger deficits to pay for that public investment.

Behind the measurements of aggregate growth, productivity, and trade figures has been, since the 1950's, a steady transformation of the American economic base. It was presumed that the United States would lead in sectors that catered to its comparative advantage as a capital intensive and technologically rich economy. In the 1950's the United States largely lost the textile and apparel industry[29] to the labor-rich countries of Asia, and moved toward the capital—and knowledge—intensive consumer electronics industry. In the 1960's American producers were gradually driven out of this sector by Asian competitors. Advanced manufacturing industries such as steel and automobiles suffered a similar fate by the 1970's as comparative advantage began to pass to foreign producers.

This evolution of economic specialization through the central dynamic of technological innovation is consistent with the "creative destruction" process articulated by Joseph Schumpeter. He argued that when technology and goods become obsolete, labor and capital leave to create new and more advanced enterprises.[30] In such situations, pain and dislocation must be subordinated togreater wealth generated in the new industries. According to this perspective,

Table 1.2
Percentages of World High Technology Exports, 1970-1989

Country	1970-1973	1979-1982	1988-1989	Change 1970-1989
United States	29.54	25.07	20.64	- 8.91
Japan	7.07	10.06	16.01	+ 8.94
Germany	16.59	14.66	12.53	- 4.08
France	7.22	8.10	6.80	- .042
UK	9.87	9.87	7.64	- 2.48

Source: World Trade Data Base. Paolo Guerrieri and Carlo Milana, "Technological Competition in High Tech Products," BRIE Working Paper 54 (Berkeley: University of California, October 1991).

the United States could tolerate a sustained loss of rust belt manufacturing to overseas competitors because it was a natural process that would unleash the resources for the wealthier, cleaner, and more value-added processes associated with high technology industries and services. America's comparative advantage in high value-added products would, theoretically, compensate for increased imports of traditional manufactured goods through the increased export of high technology goods and services. However, by the mid 1980's a range of American high technology companies suffered from intense foreign competition.

Competitiveness cannot be measured simply by a nation's ability to sell abroad and to maintain a balance of payments. Otherwise the United States must be considered an uncompetitive economic power, and this is not reflected in productivity figures, GDP growth rates, and per capita income. Moreover, export figures do not reflect sales made through foreign affiliates and subsidiaries of American companies.[31] More fundamentally, competitiveness must address the nation's ability to stay ahead technologically and commercially in those commodities and services likely to constitute a larger share of world consumption and value-added goods. In this regard, the trade record on American high technology is mixed. Table 1.2 indicates that the United States remains the largest single exporter of high technology products. However, that position has been steadily eroded since 1973. The United States share of world high technology exports declined from 29.5% (1970-1973) to 20.6% (1988-1989), a figure that is 4.5% over second-place Japan but 17% behind the EC, whose share decreased from 46.3 to 37.3% over the same period.[32]

Although America's worldwide high technology export share has steadily declined since 1970, the level of high technology exports has steadily risen. Between 1965 and 1980, the proportion of high technology products in industrial exports doubled to 50%.[33] Table 1.3 indicates that high technology exports have increased from $60 billion to $94 billion between 1981 and 1988. However, imports of high technology products have been expanding faster than exports. During the same period high technology imports increased by over 120% while exports increased by 56%. These trade figures suggest that the international and domestic market share for American high technology products is shrinking.

Table 1.3
American Trade in High Technology Products (Averaged in $billions), 1981-1994

	1981-83	1984-85	1986-87	1988-89	1990-91	1992-94
Exports	60	67	78	94	97	111
Imports	35	60	77	77	62	84
Balance	+25	+7	+1	+17	+35	+27

Department of Commerce Classification Scheme
Source: U.S. Economic Database, Federal Reserve Bank of New York

The United States has become more dependent on a broad range of imported industrial technologies. This is an unfamiliar position for a country that had long been accustomed to technological supremacy. However, the number of technologies in which the United States has a clear lead over Japan and Europe has decreased considerably over the last decade. The United States spends more on R&D than the next four countries combined, but defense has consumed the lion's share of funds for technology research and development. The Pentagon's share of all government R&D rose from 50% to 65% between 1980 and 1990, representing almost 30% of all R&D undertaken in the United States.[34] In 1991 Japan which has a GNP under two-thirds that of the United States, out-invested America in industrial R&D in absolute terms. Increasingly limited commercial spin-offs from defense R&D cannot match the results of projects funded by competitor nations that have commercialization as the primary objective.

In 1989 the Department of Defense reported that Japan had significantly overtaken the United States in six critical advanced technology areas.[35] They included microelectronics circuit design and fabrication; preparation of compound semiconductors; machine intelligence and robotics; integrated optics; superconductivity; and biotechnology materials and processing. The Council on Competitiveness has identified what it considers 94 generic technologies critical competitiveness. In a 1991 report the council indicated that the United States was losing badly, or had lost the lead, in one-third of those technologies.[36]

Table 1.4
World Trade in Electronics (Percent)

Country	1973-79	1985-88	1989
United States	28.9	19.2	18.3
Japan	9.6	23.4	24.0
EC-9	44.7	30.8	28.6
East Asian NIC's	3.1	13.4	15.1

Source: World Trade Data Base. Paolo Guerrieri and Carlo Milana, "Technological Competition in High Tech Products," BRIE Working Paper 54 (Berkeley: University of California, October 1991).

The erosion of American high technology capabilities has been notable in electronics. Table 1.4 indicates that the United States share of world trade in electronic products has decreased from 28.9% in 1973 to 1979, to 18.3% in 1988 to 1989. The American trade deficit in electronics with Japan improved slightly after the dollar devaluation in 1985 but remained as high as $19.7 billion in 1991.[37] Coincident with this is a dependence on a foreign supply of essential electronic components. A potential problem is that a dependence on electronic components, advanced materials, or machinery technologies may compromise the competitiveness of those firms that may not be assured access to the appropriate technologies in a timely fashion at a reasonable price.

The question posed at the beginning of this section was whether competitiveness was a vanishing asset. It is true that American economic power has eroded across a broad range of industries, has endured slower productivity growth, and suffers under the weight of the government deficit. At the same time, however, it is premature to conclude that the United States is in a period of fixed economic decline. There is no indication that America's economic competitors will grow at a greater rate. The United States has a comparatively low unemployment rate (7% compared to 11% in Europe), has growth rates consistent with economic rivals, leads the world in productivity, and has the technological talent and economic resources to remain a decisive world power for some time. Moreover, depending on how one reads the Council on Competitiveness report, the United States still enjoys a lead in two-thirds of those technologies perceived as critical to industrial competitiveness.

To some extent the convergence in economic capabilities among the developed nations was a logical result of the postwar system. First, the multilateral free trade regime and the technological revolution in transportation and telecommunications neutralized the advantage of American companies deriving from the size of the U.S. market. The bonds that confined markets to geographic proximity have been weakened, and many markets have largely collapsed into a single global medium. Firms once preoccupied with local competition are increasingly forced to compete with corporations from all over the world for the same share of domestic market. Second, increasing interdependence and revolutionary developments in communication technology accelerated the technological diffusion across national borders, making American technology more generally accessible. Nevertheless, the extent to which the American position has eroded across the high technology spectrum may reflect something more than the process of catching up. That many competing product and process technologies are the object of foreign industrial targeting suggest that other forces may also be shaping comparative advantage. This has nourished the perception that the United States has lost its technological dominance, not by a sudden shift in national capabilities or resources but rather as the result of a foreign government's technology program.

It is clear that the unique advantage the United States once had in high technology products and processes has diminished. The implication is that the

United States is not really moving into higher technology trade, rather it is losing the ability to produce high technology goods that meet the test of highly competitive international markets. The United States has no option but to remain a leading developer and producer of high technology products if it is to maintain its standard of living and the influence of its foreign policies. In recognition of this, the long-term competitiveness of America's high technology enterprise has become a legitimate policy concern.

RESHAPING NATIONAL SECURITY

The integration of the world economy has compromised the economic independence of all states. If economic sovereignty is defined as the power to control a full range of policy instruments, then virtually all states in the international economic order have lost some degree of economic sovereignty. Indeed the ceding of a degree of economic sovereignty is implicit in the very notion of economic interdependence. A principal concern, however, is to what extent economic interdependence and competition threaten the core values that the state is sworn (or at least obliged) to protect. Economic integration affects domestic economies in myriad ways: in access to supply of goods or markets; in levels of employment; in inflation; in return on productive factors; and in income distribution, to mention a few. Government is entrusted with the responsibility of maximizing the economic welfare of its citizens. However, those macro-policy tools at the disposal of the government are less effective since economic conditions are increasingly influenced by external factors.

The fundamental issue in any security policy is to secure that which the nation values. The objectives of national security, therefore, will almost certainly vary among nations. Security can be defined as the absence of acute threats to the minimal acceptable levels of the basic values that a people consider essential to its survival. Economic security can be defined as the absence of threat of severe deprivation of economic welfare. Economic security manifested through government policies becomes most visible when a country "consciously chooses to accept economic inefficiency to avoid becoming more vulnerable to economic impulses from abroad or when a country stresses national approaches at the expense of international integration."[38]

Security issues have always had an economic dimension. Economic considerations in security matters include economic compellence, the use of economic power to influence behavior of another state that depends on the economic resource(s) of that state; economic sufficiency in wartime, the ability to maintain the supply of vital materiel for war; and the more fundamental reality that a first-class military establishment depends on the economic power to buy it. All three of the above economic considerations are common insofar as they represent economic security as a component to traditional security founded on military power.

Security may also be measured in nonmilitary terms. Should those conditions that support and sustain core values become jeopardized, then the nation can either create conditions favorable to the preservation of those values or it can redefine those values. If we assume survival, freedom, and the pursuit of the American dream are core values, then the decline in economic welfare can be perceived as a threat to national security. When an economic threat to security appears, it is usually less tangible, less urgent, and less likely to arouse the same degree of nationalism or ethnocentrism as a traditional military threat. Nevertheless, a condition that is manifested through gradual erosion of the foundation on which security depends will eventually occasion a similar compromise to national security associated by a singular event. Any threat to the value framework of a country can be construed as a national security issue and should be treated accordingly. Economic growth and technological sophistication in Europe and Japan promoted American security interests in the postwar system by strengthening the foundation of the Western Alliance. The economic dimension of the postwar security system was a critical component in the exercise of American power, and the preponderant strength of the American economy suggested that there was no obvious need to trade off "guns for butter." Industrial and technological resources supported a large standing military and underwrote the use of commercial and technical assistance to secure Allied agreement with American security goals. This technical and commercial assistance helped integrate free world defenses, facilitated job growth in high technology, and accelerated the diffusion of American technology. However, since military might has been discounted as an emblem of power at the expense of trade and technology, there is reason to believe that more consideration of economic issues in the security context is warranted.

The FSX (Fighter Support Experimental) affair with Japan illustrated the policy conflict between military security and economic power. In the mid-1980's the Japanese government endeavored to build a new fighter jet for its "self defense" forces of the 1990's. The Japanese were more interested in the deal from a commercial point of view. It was no secret that Japan had plans to create a Japanese commercial aerospace industry and American technology could be used to facilitate that objective. The main provisions of the deal provided the Japanese with unlimited access to F-16 technology and 60% production of the plane in Japan. It was a project that the Pentagon and the State Department both supported because the arrangement conformed to their worldview of integrated military structures and geopolitical advantage. There was less inclination to examine the deal in commercial terms. The gap between trade and defense objectives militated against an institutional framework that could assess, on balance, the potentially destructive long-term implications of such agreements in economic terms.

The FSX affair highlighted the growing confusion over American security goals while underscoring the deeper policy question of whether economic interests and traditional national security concerns should be given equal weight in the formulation of American foreign policy. In October of 1989 the Defense Science Board Report entitled *Defense Industrial Cooperation with Pacific Rim Nations* advanced considerations that challenged the wisdom of official policy. The report concluded that "national security can no longer be viewed only in military terms but must include economic well being as a key component . . . therefore we must explicitly line cooperative defense technology sharing issues with economic issues, including trade balance and market access."[39] The report further articulated many of the concerns associated with sharing American technology in the interest of defense cooperation and the negative effects that could have on American industrial and technological competitiveness.

The elements of power—those qualities, conditions, and resources that lend one state (or group of states) the capacity to influence policy decisions in other states—have assumed an economic character. Those who have the capability of buying vital capital assets in foreign states or controlling the supply of vital products, technologies, commodities, or capital have the capability of exercising influence in those states or, at the very least, of threatening to exercise such influence. Unlike after World War II, the United States in the post-Cold War world is confronted with immense economic challenges that will almost certainly compromise the capacity of the nation to preserve national values and security interests in a new world order that will make different demands on national power.[40] America redefined its national security agenda after World War II and the government deployed substantial resources and energy in its victory over its Cold War adversary. In the final decade of the twentieth century, economic strength and the appropriation of technological innovation will become an increasingly vital component for the long-term security interests of the United States.

NOTES

1. Until the Korean War made rearmament possible, the United States had to reconcile post-World War II demobilization with a tiny atomic arsenal to deter the Soviets. President Eisenhower, elected toward the end of the Korean War, considered matching conventional forces with the Soviets too costly and instead opted for a policy of massive retaliation. It was a premise founded on American strategic superiority. The United States would not respond in kind but would resort to nuclear weapons to inflict unacceptable damage on the Soviets.

2. American prosperity in those years was founded on five specific strengths that included the most advanced technology, a domestic market eight times the size of its largest competitor, a highly skilled workforce, a huge supply of capital, and a sophisticated culture of business management. Michael L. Dertouzous, Richard K. Lester, and Robert M. Solow, *Made in America: Regaining the Productivity Edge* (Cambridge: MIT Press, 1989), 23-25.

3. Organization for Economic Cooperation and Development, *Economic Outlook* (Paris: Organization for Economic Cooperation and Development, 1975), 33-37.

4. For example, the trade percentage of American GNP doubled to 25% from 1960

to 1983. U.S. International Trade Administration, U.S. Department of Commerce, *U.S. Competitiveness in the International Economy* (Washington, D.C.: Government Printing Office, 1984), 16.

5. David Dollar and Edward Wolff, "Convergence of Industry Labor Productivity Among Advanced Economies, 1963-1982," *Review of Economic Statistics* 70 (November 1988): 549-558. See also Moses Abramovitz, "Rapid Growth Potential and Its Realization: The Experience of Capitalist Economies in the Postwar Period," in *Economic Growth and Resources*, Edmond Malinvaud, ed. (London: Macmillan Press, 1979), 23. David Calleo, *Beyond American Hegemony: The Future of the Western Alliance* (New York: Basic Books, 1987); Stephen Cohen and John Zysman, *Manufacturing Matters: The Myth of the Post-Industrial Economy* (New York: Basic Books, 1987); Robert Gilpin, *War and Change in World Politics* (New York: Cambridge University Press, 1981); Charles P. Kindleberger, "Dominance and Leadership in the International Economy," *International Studies Quarterly* 25, no. 2 (June 1981): 242-254.

6. Although it is true that many years of American prosperity occurred during periods of large military expenditures, and a recent decline in defense spending coincides with extensive discussion of slipping competitiveness, it is not an indication that as defense spending goes so goes national competitiveness. As discussed above, in the first several decades after World War II the United States endured little competition with the rest of the world. Large military budgets certainly support and increase aggregate economic activity, at least in the Keynesian model. However, despite spin-offs, economic competitiveness is not related to the size of military spending, and as the large defense buildup of the 1980's illustrates, government spending may for some time create a borrowed prosperity that may burden long-term growth with large fiscal deficits.

Historian Paul Kennedy observed that the United Kingdom of the late nineteenth century suffered from the burden of extensive military commitments and its drain on national resources at a time when its economic and industrial supremacy became severely challenged. Britain had led the world in productivity, output, manufactured goods, high technology goods and processes, and real incomes before its industrial competitors (notably United States and to a lesser extent Germany) surpassed them in exports, manufacturing, and high technology capabilities. He notes that this occasioned an extensive national debate concerning competitiveness and efficiency. See Paul M. Kennedy, *The Rise and Fall of the Great Powers* (New York: Random House, 1987).

7. Christopher Freeman, *Technology Policy and Economic Performance: Lessons From Japan* (London: Francis Pinter, 1987); William Lazonick, *Competitive Advantage on the Shop Floor*, (Cambridge: Harvard University Press, 1990); Michael Piore and Charles Sabel, *The Second Industrial Divide* (New York: Basic Books, 1984); Michael Prowse, "Is America in Decline," *Harvard Business Review* 70 (July/August 1992): 34-45.

8. John A. Young and others, *Global Competition: The New Reality*, The Report of the President's Commission on Industrial Competitiveness, Vol II. (Washington, D.C.: Government Printing Office, 1985), 55.

9. For example, that Mexico ran large export surpluses during the 1980's in order to meet debt obligations stemming from the debt crisis is scarcely emblematic of an economic juggernaut. The systemic imbalance in Mexico's trade and currency regime was exposed by the Mexican financial crisis of 1994, a development that required a multi-billion dollar capital infusion from the United State's Currency Stabilization Fund and the IMF to shore up the Mexican economy and government finances.

10. See World Economic Forum, *The World Competitiveness Report: 1990.* (Geneva: EMF Foundation, 1990), 10-15.

11. U.S. Department of Commerce estimated that 70% of U.S. manufacturing out-put is directly challenged by foreign competition. U.S. International Trade Administration, Department of Commerce, *United States Trade: Performance in 1987* (Washington, D.C.: Government Printing Office, 1988), 6.

12. Research indicates that Japanese firms innovate faster and more cheaply than do American firms. See Edwin Mansfield, *The Speed and Cost of Industrial Innovation in Japan and the United States*, University of Pennsylvania Working Paper (Philadelphia: University of Pennsylvania Press, 1985), 3.

13. Michael L. Dertouzous, Richard K. Lester, and Robert M. Solow, *Made in America: Regaining the Productivity Edge* (Cambridge: MIT Press, 1989); Michael Porter, *The Competitive Advantage of Nations* (New York: Free Press, 1990).

14. Robert Hayes and Jaikumar Rachandran, "Manufacturing's Crisis: New Technologies, Obsolete Organizations," *Harvard Business Review* 66 (September/October 1988), 77.

15. Jean-Claude Derian, *America's Struggle for Leadership in Technology*, translated by Severen Schaeffer (Cambridge: MIT Press, 1990), 264.

16. Since 1990 the United States has run a trade surplus in services, exporting approximately 30% more services than are imported. America's trade deficit is, therefore, a trade deficit in goods.

17. International Monetary Fund, *International Financial Statistics Yearbook 1994* (Washington, D.C.: IMF, 1994), 142-143.

18. Drugs and medicine did well while aerospace exports held relatively constant. See Organization for Economic Cooperation and Development, *Handbook of Economic Statistics* (Paris: OECD, 1991), 47.

19. Richard C. Marston, "Price Behavior in Japanese and U.S. Manufacturing," *National Bureau of Economic Research*, Working Paper no. 3364 (Cambridge: National Bureau of Economic Research, 1990). See also Ferdinand Protzman, "Why a Lower Dollar Didn't Work," *New York Times*, 1 December 1992, D1.

20. Robert McCauley and Steven Zimmer, *Explanations for International Differences in the Cost of Capital* (New York: Federal Reserve Bank of New York, 1989), 58. Cost of capital measures include the impact of tax policy, depreciation allowances, interest rates, and risk. Interest rates remain higher in the United States than in Japan. See "Economic and Financial Indicators," *The Economist,* 26 August 1993.

Reasons for a higher cost of capital for American companies are varied, but fundamentally it can be attributed to an enormous budget deficit and low national savings rates. A low savings rate and the need to finance deficit spending squeezes liquidity in the system, thereby raising interest rates. The average savings rate in the United States for disposable income is 4.6%; in Japan it is almost 16%. "Personal Savings Rise," *The Japan Economic Journal*, 15 September 1990, 89. In the absence of pronounced budget deficits squeezing available capital, supply and demand dictates a cheaper supply of capital.

21. Hurdle rates (required rates of return) on R&D projects with a ten-year horizon were 8.7% Japan, 20.3% United States, 14.8% Germany. For plant and machinery with a 20 year horizon the hurdle rates were 7.0% Germany, 7.2% Japan, and 11.2% United States. Simon Holberton, "The Differing Costs of Capital," *The Financial Times*, 1 June 1990, 16.

22. Ira Magaziner, "Growing Our Economy," *Business & Economics Review* 39 (April/June, 1993), 8-12. Magaziner states that because productivity is improving less rapidly, the real wage the average American can earn for an hour of work has gone down 17% since 1973.

23. American manufacturing productivity has risen more rapidly than in any other country but Japan since 1979. See International Monetary Fund, *International Financial Statistics Yearbook 1988* (Washington, D.C.: International Monetary Fund, 1988), cited in Michael Porter, *The Competitive Advantage of Nations* (New York: The Free Press, 1990), 6. These gains may represent a one-time surge. First, productivity gains were reached by terminating inefficient plants. Second, the Department of Commerce has recognized several flaws in the way it gathers statistics on productivity and the impressive figures in manufacturing productivity are considered inflated. See U.S. Department of Commerce, "Gross Product by Industry: Comments on Recent Criticisms," *Survey of Current Business* (July 1988), 132-133. For a more extensive treatment of manufacturing productivity see Michael L. Dertouzous, Richard K. Lester, and Robert M. Solow, *Made in America: Regaining the Productivity Edge* (Cambridge: MIT Press, 1989).

24. Productivity increased in the United States at a fantastic rate from 1945 to 1967 of 3.2% annually. At that rate, incomes will double every 21 years. Productivity slowed in the United States from 1967 to 1979, reflecting the era of stagflation and the adjustment to expensive oil. Commercial competitors of Europe and Japan experienced a similar downturn in those years but outperformed the United States through the 1980's. International Monetary Fund, *International Financial Statistics Yearbook 1991* (Washington, D.C.: International Monetary Fund, 1991), 85.

Annual percentage growth in labor productivity of Japan and Germany has doubled and sometimes quadrupled the rate in the United States which has hovered around .5 to 1% a year from 1967 to 1987. Organization for Economic Cooperation and Development, *Industrial policy in OECD Countries: Annual Review* (Paris: OECD, 1990), 92.

25. The math test scores of the top 1% of American high school seniors would place them in the 50th percentile in Japan. In science, American 10-year-olds place eighth in the world, American 13-year-olds place 13th. Thomas D. Cabot, "Is American Education Competitive?" *Harvard Magazine*, Spring 1986, 14. See also the average test scores on international mathematics furnished by the Educational Testing Service.

26. This is not the result of larger high technology industrial sectors in Japan and America, as Table 1.1 indicates that Europe maintains a larger share of world high technology exports.

27. David Alan Aschauer, "Infrastructure: America's Third Deficit," *Challenge* (March/April 1991), 42.

28. David Alan Aschauer, *Public Investment and Private Sector Growth* (Washington, D.C.: Economic Policy Institute, 1990), 17.

29. The multifibre agreement governing textiles has encouraged the retention of some domestic textile and apparel industry, however.

30. Joseph Schumpeter, *Capitalism, Socialism, and Democracy* (New York: Harper Brothers, 1942), 169; Joseph Schumpeter, *Business Cycles: A Theoretical, Historical, and Statistical Analysis of the Capitalist Process* (New York: McGraw-Hill, 1939).

31. According to one account, such sales exceed the value of all our exports, and therefore balance of payment figures do not entirely reflect the competitive position of American firms. See Alfred Balk, *The Myth of American Eclipse: The New Global Age*

(New Brunswick: Transaction Publishers, 1990), 52.

32. Not including Greece, Portugal, or Spain.

33. John A. Young and others, *Global Competition: The New Reality*. The Report of the President's Commission on Industrial Competitiveness, Vol. II. (Washington, D.C.: Government Printing Office, 1985), 76.

34. Since 1990, the military R&D share has slowly decreased as the share of civilian R&D has risen.Organization for Economic Cooperation and Development, Handbook of Economic Statistics (Paris: OECD, 1991), 67. Robert DeGrasse, "The Military and Semiconductors," in *The Militarization of High Technology*, John Tirman, ed. (Cambridge: Ballinger Publishing Company, 1984), 79. For non-defense R&D as a percentage of GNP in 1989 Japan spent 3.0%, Germany 2.8%, and the United States 1.9%. U.S. National Science Board, *Science and Engineering Indicators* (Washington, D.C.: National Science Board, 1992), 108-110.

35. U.S. Department of Defense, *Critical Technologies Plan for the Committees on Armed Services* (Washington, D.C.: Government Printing Office, 1989), 11.

36. U.S. Council on Competitiveness, *Gaining New Ground: Technology Priorities for America* (Washington, D.C.: Government Printing Office, 1991). See also U.S. Department of Commerce, *Emerging Technologies* (Washington, D.C.: Government Printing Office, 1990).

37. American Electronics Association, cited in Laura D'Andrea Tyson, *Who's Bashing Whom: Trade Conflict in High Technology Industries* (Washington, D.C.: Institute for International Economics, 1992), 27.

38. Klaus Knorr, "Economic Interdependence and National Security," in *Economic Issues and National Security*, Klaus Knorr, ed. (Lawrence: Allen Press, 1977), 14.

39. Malcolm Currie and others, *Defense Industrial Cooperation with Pacific Rim Nations* (Washington, D.C.: Office of Under Secretary of Defense for Acquisition, 1989), 10.

40. The military is not certain how to redefine America's security policy. The Pentagon recently released a study outlining American security goals in the post-Cold War world. It was a plan premised on thwarting rival economic powers with a smothering strategy. The position paper was roundly criticized by President Bush, other government agencies, and foreign governments. It was summarily abandoned.

2

TECHNOLOGY

STRATEGIC INDUSTRY

The preceding chapter outlined various elements and measurements that determine, to a greater or lesser extent, the level of economic performance. On one level, an exploration of these points can reflect the relative strengths and weaknesses of an economy. On another, such indicators do not consider the systemic and structural aspects of economic activity that have an impact on the general capacity of a nation to maintain a rising standard of living or the ability of that nation to protect from foreign threat that same economy. Those industries that add to this economic network may have a special value, a strategic dimension, but orthodox economic theory leaves little room for the concept of strategic industries. Goods and services are produced according to laws of comparative advantage, and what is produced matters less than how it is produced. It does not matter if one produces high technology or low technology goods. What is important is how efficiently it is produced and how it adds to productivity. In this world a hundred dollars worth of potato chips is as valuable to society as a hundred dollars worth of computer chips.

The literature in support of industrial policy has been cast in an environment in which markets do not function perfectly, and in industries whose characteristics imply a departure from the orthodox paradigm of pure competition. This study does not presume to be the first to reveal underlying weaknesses of the neoclassical paradigm. The limits and flaws of the assumptions and predictive capacity in the model are readily acknowledged by most economists. One of the main reasons why the orthodox model continues to exert such influence is because there is no satisfactory alternative that has the same degree of overall rigor. However, market distortions that include substantial economies of scale, steep learning curves, positive externalities, large R&D requirements, and large fixed costs of entry indicate that a departure from the orthodox paradigm of efficient free markets is becoming the norm rather than the exception.[1] Externalities arise when the actions of one party affect others in ways not mediated by the marketplace, and in the case of a positive externality, the value of the outputs exceeds the value of the inputs, resulting in a net surplus for the economy. The externality may be manifested in knowledge spillovers, learning by doing, and linkages that promote competitiveness in related sectors of the economy. These factors are incompatible with the efficient free-market model and provide an economic rationale for government intervention, one that contributes to the net welfare of the economy.[2] What distinguishes government intervention today from nineteenth century mercantilism is that there is an economic case for intervention that will not merely shift economic assets but will achieve a net profit or surplus for the economy. Mercantilism embodied strategic objectives, but orthodox theory held that government intervention would be welfare reducing.

There is no airtight definition of a strategic industry. Economists at different times and of varying persuasions have emphasized different characteristics of what it means to be strategic. The term *strategic industry* has typically referred to an industry most important to the successful conduct of war and the maintenance of national security. Identifying an industry as strategic for defense is relatively straightforward, as one calculates the role that an industry has in contributing to the arsenal, whether it be tanks or ball bearings. It is more complicated to determine strategic value in a commercial sense, because it is based on intangible values not captured by market indications of worth. Yet the requirements are similar in that a strategic industry is one of great importance within an integrated framework.

At first glance industries that are the primary cause of economic growth may appear to have a strategic dimension.[3] Industries that are keystones to other sectors through upstream and downstream linkages do occupy a central role in economic development. Such sectors may be considered strategic insofar as their actions generate economic growth in those industries through intersectoral linkages. However, this condition alone is insufficient in identifying a strategic activity. Many industries may satisfy the above condition while not being strategic. Agriculture, education, medical services, and financial services are all important activities but are not necessarily strategic. People must eat but that does not make agriculture a strategic industry in the context of this discussion. Education may improve productivity through the systematic application of knowledge to generate more and better output from inputs, yet it is not fundamental to national or economic security. Medical services keep people healthy and thereby contribute to improved human resources but, like education, are not *a priori* a prime determinant of productivity or security. Cigarettes employ large amounts of tobacco, paper, and chemicals yet few would argue that such linkages justify promotion of the industry. Determination of an economic activity as strategic depends in part on the characteristics of the economy and the society in which such activity operates. Yet a common element in identifying strategic sectors, and the most important for our purposes, is its putative importance in national security considerations. Since the ability to preserve national values is increasingly a function of economic vitality, we may add that a strategic industry is one that is critical to the economic interests of the country.

A strategic industry is one that is essential to the economic and national security interests of the state and one that engages in an activity that affects the national economy with critical forward and backward linkages through the existence of positive externalities. The American government defines an industry as strategic if it meets two criteria: one is that progress in the underlying technology of the industry forms the foundation for productivity growth; and the other is that the industry is linked to other industries such that technological progress in the former affects the rate of progress in the latter.[4] In this capacity, the strategic industry provides network externalities and is an important agent in economic growth and the elevation of national welfare. By extension, dependence on foreign suppliers in that industry increases the likelihood of strategic withholding, and this can have negative consequences for national welfare.

DEFINITION OF HIGH TECHNOLOGY

In advanced economies high technology industries enjoy varying degrees of support. Many are leading-edge growth sectors that survive on the merits of innovation and rely heavily on the application of new science-based technologies. High technology industries share common features such as large investment requirements in R&D, economies of scale, learning by doing, and product cycles compressed by continual innovation. Table 2.1 identifies three similar classification schemes for high technology industries. All three use quantitative indicators to classify a high technology industry. The Department of Commerce defines a high technology industry as one having a high proportion of R&D costs as a percentage of sales and one that employs an above-average number of scientists and engineers in its workforce.[5] The OECD (Organization for Economic Cooperation and Development) employs a more narrow definition, one that does not include industries that employ sophisticated production techniques in the high technology category.[6] Like the OECD and Commerce Department, the Guerrieri-Milana classification scheme identifies high technology industries as those with above-average R&D spending and a high ratio of scientists and engineers. The Guerrieri-Milana classification scheme refines the OECD scheme by breaking down data on world trade flows and base classification decisions In a more subjective approach.[7]

High technology industries have been categorized by the above standards, but, as in the case of defining a strategic industry, there is no precise criterion by which to classify a high technology industry. There is broad consensus on those core industries that make up the heart of the high technology economy.

Table 2.1
Classification Schemes for High Technology Industries

U.S. Department of Commerce Standard Industrial Classification
-Guided Missile and Spacecraft
-Communication equipment and electronic components
-Aircraft and parts
-Office, computing, and accounting machines
-Drugs and medicines
-Industrial organic chemicals, plastic materials, synthetic resins, fibers
-Ordnance and accessories

Organization for Economic Cooperation and Development
-Drugs and medicines
-Office machinery and computers
-Electrical machinery
-Electronic components
-Aerospace
-Scientific instruments

Guerrieri and Milana
-Chemicals
-Pharmaceuticals
-Electronic components and office machines
-Data processing
-Telecommunications
-Aircraft and parts
-Scientific instruments
-Power-generating machinery

Source: Laura D'Andrea Tyson, *Who's Bashing Whom: Trade Conflict in High Technology Industries* (Washington D.C.: Institute for International Economics, 1992).

Industries such as aerospace, semiconductors, drugs and medicine, microprocessors, telecommunications, and computers are all commonly regarded as high technology enterprises[8] (See Table 2.1). Beyond these specific industries (and even subindustries of the above) there is less agreement. Industries that are users of high technology are not considered *a priori* as a high technology enterprise.[9] For example, a case could be made that the automobile industry should be considered as high technology given the large degree of R&D invested in the production of new vehicles. However, the definition of high technology industry as employed by this work is one that is largely, though not exclusively,[10] an intermediate product that is employed in the production process and as capital or intermediate product that is diffused more widely throughout the economy. Using this criterion the aerospace industry may, at first glance, resemble the automobile sector more than an industry such as semiconductors. However, given the putative importance of aerospace in the maintenance of national security in a traditional military context, it does merit greater consideration when weighing issues relating to economic security, and it is for this reason that it commands such attention among policy makers and academics. The exact membership of enterprises in the high technology category is thus subject to some variation, particularly in identifying fringe industries that may or may not have the sophistication to meet the above tests. High technology industry is not necessarily strategic. But economic activity that meets the strategic definition and has a transformative dimension important for intersectoral linkages is typically found in the high technology sector.

THE ECONOMIC IMPORTANCE OF HIGH TECHNOLOGY INDUSTRIES

The report of the President's Commission on Industrial Competitiveness, *Global Competition: The New Reality*, confirmed the primary importance of technology in preserving American strength. The report submits that "fueled by R&D, technological innovation is vital to America's future and it is the key to productivity advances. Over the past 50 years it has been the most important generator of productivity growth, far surpassing the contributions of capital, labor, and economies of scale. . . . It will be the primary factor in determining our economic vitality."[11] Productivity is a widely accepted measurement of economic progress as it reflects increased efficiencies and higher output given a set of existing inputs. Although technological innovation is but one of several important elements to productivity, it is perhaps the most essential. Roughly half of the productivity gains can be traced to technological advances that are reflected in improved manufacturing methods, materials, and machinery.[12] At the most elementary level competitiveness is achieved by maximizing output with existing inputs, and for this purpose alone technology is indispensable.

R&D and Positive Externalities

High technology industries account for 20% of America's manufacturing output and 24% of its manufacturing value-added, but nearly 60% of national industrial R&D.[13] There is consensus that technological innovation through research and development is one of the most important sources of growth in national income and productivity. There is less agreement on the specific level of growth provided by national R&D.[14] Although the exact rate of return on R&D is subject to debate, it is clear that technological innovation is a major source of economic growth. Research and development is a vital source of technological progress. However, the intrinsic importance of some forms of R&D does not necessarily mean that the federal government should assume a central role in shaping the commercial research agenda. There has been an enormous amount of private-sector R&D. This suggests that in the vast majority of cases a strategic industrial policy is not necessary. The case for federal promotion of commercial R&D requires more than an understanding that R&D is good for the nation. It requires an assessment of whether the private sector, in response to market forces, will underinvest in R&D. By extension, it is important to assess where the combination of market failure and external benefits together provide for a reasonable justification of government promotion.

In a well-functioning capitalist market economy with capital markets, banks, venture capitalists, and other financial institutions, one would expect that all lucrative possibilities would be identified and exploited accordingly. However, there is broad agreement that there are occasions when the private investment that drives capitalism neglects or underinvests in certain activities.[15] This situation is most evident in research and development, and this is one reason why R&D schemes tend to be the favorite of industrial policy advocates.

Insufficient investment in R&D may be attributed to the two problems of appropriation and asymmetries of information in capital markets. Several studies have indicated that public returns on innovation are greater than private returns, and one estimates that society's overall return on R&D is at least twice the private return.[16] Private enterprises that do invest heavily in R&D rarely capture the full value of the investment. Despite patent and copyright protection much of the reward from private innovation is enjoyed by imitators and consumers. More fundamentally, much of the reward for innovation is captured as improvements to the industrial framework which many other enterprises are linked to and depend on. Questions of appropriation may therefore diminish the incentive of a firm to invest in R&D, despite indications that society may benefit by an amount that more than justifies its expenditure. It is this incongruity of risk to reward that may create a condition of underinvestment.[17]

Since many firms are unable to generate internally the capital required to finance new ventures, capital markets assume an important role in the development of new R&D activities. To a greater or lesser degree, all business investment has some element of risk. However, additional risks are evident in R&D because a maximum return may be generated only if the firm is the first to apply it. Moreover, the evaluation of that risk can be more difficult, because in many cases it is an endeavor that establishes new directions with uncertain precedents. Although borrowers will have a better sense of the inherent risk of the project, conveying that knowledge to lenders can be problematic, and the consequent uncertainty of R&D investment promotes a trend to overestimate the risk of new activities, thereby raising the cost of capital formation.[18] Higher capital costs make investments more expensive and therefore reduce the overall level of investment.

The intersectoral linkages and positive externalities associated with the high technology enterprise are fundamental reasons why this economic activity is of special relevance to the economic power and preeminence of a nation. It is not so much a question of nurturing national champions as it is of sustaining a vital industrial network where certain industries become instrumental for long-term productivity growth and economic competitiveness.[19] It should be noted that direct evidence that high technology industries generate returns to society above their direct returns is elusive. Technological spillovers are difficult to measure, and since externalities have no price, the level of benefit to the national economy is uncertain. The difficulty in establishing a direct causal relationship should not be surprising, however. Externalities and spillovers, as their definitions would suggest, are beyond the effective predictive ability of the economics profession. Yet it is reasonable to assume that external benefits can be generated from technological innovation and dispersed through intersectoral linkages.

Employment & Wages

High technology industries have the highest productivity and compensation. A 1989 study indicated that, on average, the annual compensation in high technology industries was 22% higher than the average for all manufacturing industries.[20] Although this is a normal reflection of higher education, skills, and capital investment, high technology industries can be regarded as strategic when one considers that an advanced economy must be competitive in at least some high technology products to provide for a rising living standard and the ability to meet international economic and/or military obligations. High wages may be sustained through high productivity, but can be preserved only if a high-wage country is capable of creating goods that low-wage countries are technically incapable of producing. The product life cycle suggests that low-wage countries will gain comparative advantage once a technology becomes standardized. The high-wage, technologically advanced economy must therefore engage in a continual process of innovation in order to maintain its privileged position. Investment in R&D becomes a strategic necessity.

Technology and Manufacturing

The postindustrial theme articulated by Daniel Bell in the mid-1970's and embraced in popular culture by the start of the 1980's suggested that service industries harnessed the comparative advantage in intellectual and technological capital of an advanced economy and was, therefore, the ultimate form of economic activity. Accordingly, manufacturing is not intrinsically more valuable and it should emigrate to countries that have comparative advantages in industrial processes and labor. According to this position, high technology activities are not particularly special beyond market-indicated value.

The presumption that the United States would succeed on a diet of clean, high technology manufacturing and services may have acted as an elixir as comparative advantage in industry after industry slipped overseas. With 70% of the national workforce in service employment compared to under 20% in direct manufacturing, and with manufacturing accounting for only 25% of GDP,[21] postindustrialists suggested it was already happening. Since 1980, 21 million jobs were created in the service sector while employment in manufacturing declined by over 10%.[22]

To adopt the premise of a postindustrial society is to view the world as fundamentally benign, where economic and political realities are divorced and the supply of critical goods is assured by friendly commercial rivals. It is a nonstrategic orientation. Ignoring distinctions in value-added, in this world potato chips are indeed as valuable as computer chips. However, the logic of a postindustrial society suffers notable flaws. Economic activity rarely stands in isolation, as actions in one arena have ramifications in another. The traditional manufacturing industries are important clients of the high technology sunrise sectors that largely make producer goods: equipment, components, subsystems, machinery, and advanced materials to be used in the production of final consumer products. Failure to retain a traditional manufacturing capability will, therefore, undermine advanced technology sectors by eroding their customer base. Advanced economies have a comparative advantage in high technology because of the peculiar set of conditions that set them apart from developing countries, and it is worth exploiting that advantage.

The link between manufacturing and services is important to national competitive advantage in service industries. While manufacturing is less important in its direct employment capacity, it drives service industry growth through the buyer-supplier relationship, through services tied to the sale of manufactured goods, and through manufactured goods tied to the sale of services. Much of what is of value in services is tied to one's expertise in producing things in a more innovative or efficient fashion. Service industries such as insurance houses and consulting firms are at a competitive disadvantage when manufacturing emigrates (service links upstream such as retailing and advertising are not as affected). Factories and firms requiring such services are going to source it locally because companies closer to the source of the client are more attuned to financing requirements; are in a better position to offer timely service; and are subject to the same business, cultural, and macroeconomic climate of the client.

The need to manufacture is not only an economic necessity, but an imperative for national security. The Pentagon relies heavily on domestic industry to source the materials necessary in times of war. It would be politically and militarily untenable to rely on foreign technology and material for national defense requirements. The question then becomes not whether the United States should maintain a manufacturing capacity but what type of capability will be retained. Given the challenge to American dominance in high technology, the erosion of our presence in traditional smokestack manufacturing industries, and the systemic problems associated with trade and budget deficits, a more structuralist view in the debate over national competitiveness has emerged. Stephen Cohen and John Zysman observe:

> A decline of manufacturing presages a decline in associated high wage services. . . . Links that promote ongoing market adaptation and technological innovation mean that advantage in a national economy is embodied not simply in the capacities of individual firms but in the web of interconnections that establishes possibilities for all. Further since making links within the national economy creates real advantages and speeds the development of the most advanced technologies and the applications of these new

possibilities even to traditional industries. . . . Strategic sectors are those whose products and processes alter or transform the goods and production arrangements throughout the economy.[23]

The Science Policy and Technology Committee of the OECD has also emphasized these structural aspects of economic power, stating that "the competitiveness of a national economy is more than the simple outcome of the collective competitiveness of its firms; there are many ways in which the features and performance of a domestic economy viewed as an entity with characteristics of its own will in turn affect the competitiveness of firms."[24]

Once a more structuralist approach to examining domestic economic activity is introduced it becomes increasingly clear that an economy cannot in isolation generate economic success through a reliance on services. Indeed, the matrix of structural and operating arrangements has powerful implications for economic strength. A country cannot expect to be a world economic power unless it nourishes the industrial network on which national power is based. Mature manufacturing industries, high technology industries, and service industries all have an important part in contributing to economic health and all are mutually dependent, either directly or indirectly, on the active presence of the other. In addition, technological development depends critically on maintaining connections among the producing firms, suppliers, and customers.

Once we accept the premise that manufacturing is a fundamental component of economic power and national wealth, then the importance of technology tied to its economic mission becomes apparent. To be competitive one has to be efficient. Technologies are molded by the needs and relationships that exist in the countries from which they emerge. The technological competitiveness of firms depends, in large measure, on the external economies associated with the national system of innovation. Indeed a principal reason high technology has been made an issue in connection with industrial policy is because of its systemic character. High technology industries such as semiconductors and advanced materials provide intermediate building blocks to increase the quality and productivity in production. Since labor costs are a dwindling percentage of the value-added, real value resides in the design and technology involved in producing the good. H. W. Coover emphasizes the technological framework and manufacturing nexus directly:

In the future, new technologies will require much closer integration of R&D with engineering, manufacturing, and marketing, and thus also with corporate and business unit strategic planning. In the future, virtually every major industry in the developed world, whether now classified as high tech or smokestack, must in fact be high tech to succeed in international competition. Industries will survive and thrive only by integrating advanced information, manufacturing, and computing technologies into their designs, products, and processes, and only through high levels of innovation, quality, and reliability.[25]

Orthodox economic theory does not adequately account for the interdependent structural relations that link the national productive system. Traditional concepts of comparative advantage merely inform us that factors of production should move to sectors that use factors most productively, and that declining sectors can be viewed in isolation. However, inadequate technological capital can place an advanced economy at a competitive disadvantage. The notion that the "competitive nation" concept does not have much meaning for economic prosperity ignores the economies of scale and externalities captured by a system of high technology product and system development, national laboratories, and research consortia. It can readily be argued that economic activity that adds to this technology infrastructure may be regarded as particularly beneficial to national economic development. Such sectors that do prove more advantageous in the long run would not be predicted by traditional factor employment theory.

Military and Technology

High technology is crucial to a strong defense as well as to a strong economy. The International Trade Administration pointed out:

> Advanced technology products and the industries that supply and develop them form a critical foundation for our defense capability. . . . High technology industries take on a strategic importance and the maintenance and protection of a broad U.S. technological base is vital to national security. . . . Failure to maintain technological leadership could lead to adverse consequences for the United States.[26]

Virtually every modern tool of war depends critically on high technology goods, from new materials and composites to advanced microelectronics. Throughout the Cold War American high technology was a cornerstone that informed our defense strategy. Quality smart weaponry has been the trump card in American defense and was employed to offset the numerical advantage enjoyed by the Soviet Union. Military success is, therefore, a function not merely of access to critical technology but access to the best technology. Should the United States lose the ability to provide for the broad technological requirements of the military establishment, then security risks will certainly increase.

The importance of high technology has never been lost on the Department of Defense. However, despite this long-term commitment to high technology, American industries crucial to the broad technology requirements of the military are more vulnerable than ever before. American aerospace is being effectively challenged by Airbus, and the Japanese have made no secret of their plans to develop their own aerospace industry. Semiconductors, composite materials, and sensing devices are all under intense competition from Japan. The Pentagon requires strong domestic industries to maintain the technological edge. At any time, and especially in an era of diminished defense spending, the Pentagon cannot expect defense contracts alone to ensure the vitality of crucial high technology industries. The Pentagon depends on satisfying most of its procurement needs with what are called COTS (civilian off-the-shelf technologies). COTS represent technologies provided by commercial operations and not by specific Pentagon R&D programs. In recent years many U.S. producers providing such goods have been undermined by foreign competition. The Pentagon views this with alarm.[27] Increasingly, the Department of Defense has become compelled, however reluctantly, to include foreign suppliers and producers in the procurement process. It is possible that timely and continued access to vital technologies may not be guaranteed if the United States fails to be commercially successful in those very technologies. The production of boots and bullets for the military fulfills a need, but it is substantially less important to national defense than advanced microelectronics, aerospace, and associated high technology goods.

The potential risk that emigrating high technology poses in the military context is one of sourcing. In time of war, would the Department of Defense have assured access to those critical technologies to maintain the war machine? In theory any and all industries that support a war effort may be considered strategic if those industries offer goods unavailable elsewhere. One could readily argue that since the industrial economies of Europe and Japan are our political allies, and will remain so for the foreseeable future, then sourcing goods and technology for defense would not be severely compromised. A caveat to this logic is the possibility that our friends today may be our enemies tomorrow. Relations among the Western nations are changing. Without any clear post-Cold War threat and the ascendancy of potentially divisive economic issues,[28] the possibility exists that control over key high technology systems could be used to influence national behavior. The question then becomes one of sourcing in peacetime and what potential pitfalls might be inherent in a procurement program that relies, to a greater or lesser extent, on foreign technology.

The Office of Technology Assessment argues that predicting and keeping current on technological progress and applications would, at best, be extremely difficult with the above scenario.[29] Being removed from the mainstream of technological innovation owing to geography, language, or interaction would eventually make it difficult to maintain the state-of-the-art technology required by the military complex. The Office of Technology Assessment also reported that foreign suppliers of computer parts and machine tools sold to U.S. companies embodied outdated technology that is two to three years behind that offered to their domestic customers.[30] The Defense Production Act mandates that should the military require a part, component, technology, or whatever else it deems necessary in an emergency situation, the supplier must meet these demands regardless of their effect from a commercial perspective. This would not apply to overseas suppliers. Foreign sourcing for military or commercial technology thus becomes one of questionable access and delay.[31]

Spin-off to Spin-on

Many industries such as aerospace, semiconductors, computers, numerically controlled machine tools, lasers, and nuclear power attribute their early success to defense efforts in the 1950's and 1960's.[32] Initial applications were developed for and procured by the military and later diffused into commercial use. In most cases this has fostered technology leadership in both defense and civilian capacities. In semiconductors government R&D helped set the functional characteristics of the new technologies, R&D funds accelerated the development, and military procurement provided a guaranteed market that hastened innovation and the move down the learning curve.[33] By providing a guaranteed market for integrated circuits and transistors when the industry was in its infancy, the government absorbed the high cost of innovation at a time when prices were too high for widespread civilian application.

Spin-offs occasioned by military procurement continue to offer advantages to the commercial sector, but with a declining frequency[34] because modern military R&D has become more sophisticated and mission oriented. What the military often needs in a semiconductor chip—unsurpassed performance in potentially hostile conditions—is not what the civilian industry needs—low cost and reliability. The subsidization of esoteric military technologies such as nuclear-resistant microchips, stealth bombers, night-vision devices, radar jamming equipment, and missile guidance systems offer little commercial utility.[35] Since the military technology development trajectory no longer provides extensive technological spin-offs, it cannot be a source of American technological preeminence. The Young Commission went further, concluding that "the military in the 1980's was a net user of the pool of civilian research and no longer a net adder to it. . . . The days of an automatic spillover to the domestic economy from military and space-aimed R&D were past; the nation's stock of scientific and technological talent was a resource from which the military was now subtracting."[36]

An emerging technology development trajectory in which technology diffuses from civilian to defense has been coined "spin-on." Many militarily relevant subsystem components, machinery, and materials technologies are driven by high-volume commercial applications that produce leading-edge sophistication compatible with military specifications. The electronics industry is the clearest illustration of this trend. High-volume electronics is driving the development, costs, quality, and manufacture of technological inputs critical to computing, communication, weapons, and industrial electronics. Products such as small televisions, liquid crystal displays, optical disk mass storage systems, and portable telephones have a wealth of silicon chip technology, and have been a principal factor motivating Japan's quest for supremacy in the semiconductor industry.[37] These products also contain optoelectronic components, LCD shutters, scanners and filters, and semiconductor lasers, all of which are, or in the stage of becoming, important components in military technologies and systems. Liquid crystal display technology developed in small flat panel screens is also finding its way into military applications. The growing sophistication of consumer and industrial products approaches and in some cases surpasses the requirements for military applications.[38]

The new technology development trajectory suggests that the old system of military technology development cannot be expected to relieve competitive challenges in high technology industries. At the same time, past reliance on military contracts during the Cold War nourished a domestic-industrial enclave that is decidedly different from a commercial world, which more than ever is the primary determinant of profit and loss, of success and failure. The major goal of American defense and industrial policy has been to maintain the industrial basis for military strength in terms of quality and quantity, but not of price. The Pentagon is not indifferent to cost, but it is clearly secondary to performance. By contrast, expense is a central concern in the civilian enterprise and price-sensitive consumer applications demand that the unit cost of underlying technological components be as low as possible. At a time when many relevant military technologies are driven by commercial competition, the Pentagon cannot be a patron for its wide-ranging technological needs, even those central for security systems, and must depend upon American industry being commercially competitive.

REPRISE

Technology has been shown to be a fundamental ingredient in productivity, in economic growth, and in a strong military establishment. The importance of technology attached to its military mission will remain critical to American security. However, the rising intensity of global economic competition suggests that the commercial dimension of high technology and the associated economic benefits are no less important in protecting the long-term security interests of the nation. It is no surprise, therefore, that the debate concerning government intervention in selected industries has gathered momentum in light of foreign targeting of technology-intensive industries linked to future economic prosperity.[39]

The results of strategic competition may be heavily influenced, positively or negatively, by government policy. Competition in these industries does not conform to models of perfect competition. These firms typically operate in imperfectly competitive markets with economies of scale, learning by doing, and R&D capabilities defining comparative advantage.[40] It is a man-made comparative advantage. Maintaining that advantage can be costly, however, because the cost of innovation has risen sharply over the past several years owing to the increasing complexity of technology and the efforts required to extend the present technological frontier. It is becoming increasingly difficult for a single firm to marshal sufficient resources and expertise to promote innovation from research to development.[41] If government promotion can augment critical resources, the possibility exists for government to tilt the terms of competition in favor of a domestic industry.

Setting the appropriate guidelines to steer government intervention remains a daunting task, but the explicit cultivation of selective economic sectors for national power has historical precedent.[42] The strategic importance of any given industry is bound to change over time. At present, the strategic industries of the advanced trading partners are going to be technology intensive. They are key components of economic growth, trade performance, and national security, and they assume a central role in shaping future economic progress. By extension, those who have the critical technologies can influence the behavior of those states that require those technologies. Stephen Cohen and John Zysman queried:

> Will U.S. producers of computers be able to stay ahead of Hitachi if they depend on Hitachi for semiconductors? Will U.S. semiconductor makers be able to stay ahead of—or keep up with—Fujitsu if they have to depend upon Fujitsu for new production equipment? Can General Motors ever surpass Toyota in productivity if it relies on Japanese production equipment and control systems for its productivity gains?[43]

Semiconductors are the essential framework of the information society and meet our criteria of a strategic industry. Semiconductors generate technological spillovers that impact upon the competitiveness of downstream products and production processes and are indispensable components in the exercise of power in military systems and commercial technology. The National Advisory Committee on Semiconductors stated in 1989 that "today's $50 billion world chip industry leverages a $750 billion global market in electronics products and 2.6 million jobs in the United States."[44] In 1993 the world semiconductor market had grown to $77 billion, and according to the Semiconductor Industry Association, it was over $120 billion by 1995. Semiconductors are at the beginning of a decisive industrial chain forming the basis of innovation in all electronic applications. Establishing a position in this industry can give a firm or a nation a dominant position in a stream of product and process innovation.

Semiconductors are also critical components and technology drivers of every sophisticated weapons and communication system used in the conduct or deterrence of war. In 1987 the Defense Science Board reported that "U.S. military forces depended heavily on technological superiority in electronics; that semiconductors were the key to such leadership; that competitive high-volume production was the key to semiconductor advancement; and that this depended on strength in commercial markets. . . . The semiconductor industry is critical to national defense."[45] Similarly, the National Research Council submitted that the semiconductor industry is strategic given its "pivotal position in the system of institutions and practices underlying America's capacity to innovate—(employment, manufacturing infrastructure, R&D, etc.)—the case for semiconductors rests on its central importance in maintaining America's capacity to innovate and commercialize technology and all that this implies for military, industrial and technological leadership."[46] International competition in the semiconductor industry is a microcosm of the wider economic battle being waged. The stakes are economic prosperity and with it the international hierarchy of wealth and power. The National Research Council concluded in *Race for the New Frontier* that: "the U.S. advanced technology enterprise has been undervalued in the past in the national scheme of priorities and must be held as one of the country's most valued objectives."[47] In the long run, failure to remain competitive in the high technology enterprise will adversely affect our general competitive posture and undermine our ability to provide for the security of the nation.

NOTES

1. Paul Krugman, ed., *Strategic Trade Policy and the New International Economics* (Cambridge: MIT Press, 1986); Elhanan Helphman and Paul Krugman, *Market Structure and Foreign Trade: Increasing Returns, Imperfect Competition, and the International Economy* (Cambridge: MIT University Press, 1985).

2. Alex Jacquemin, *The New Industrial Organization: Market Forces and Strategic Behavior* (Cambridge: MIT Press, 1987); Bruce Scott, "National Strategies: Key to International Competition," in *U.S. Competitiveness in the World Economy*, Bruce Scott and George Lodge, eds. (New York: Norton, 1987).

3. Julian Gresser, *High Technology and Japanese Industrial Policy: A Strategy for U.S. Policymakers*, Report prepared for the subcommittee on Trade and Committee on Ways and Means, U.S. House of Representatives (Washington, D.C.: Government Printing Office, 1980), 243-244.

4. U.S. Congressional Budget Office, *Federal Financial Support for High Technology Industries* (Washington, D.C.: Government Printing Office, June 1985), 77-78.

5. U.S. International Trade Administration, Department of Commerce, *An Assessment of US Competitiveness in High Technology Industries* (Washington, D.C.: Department of Commerce, 1984), 4.

6. T. A. Abbot, "Measuring High Technology Trade," *Journal of Economic and Social Measurement* 17 (1991), 17-44.

7. Paolo Guerrieri and Carlo Milana, "Technological and Trade Competition in High Tech Products," *BRIE Working Papers* 54 (Berkeley: University of California, Berkeley, 1991).

8. Romesh Diwan and Chandana Chakraborty, eds., *High Technology and International Competitiveness* (New York: Praeger, 1991), 22.

9. In this capacity, industries including VCRs and consumer electronics are also not considered high technology but rather the beneficiaries of high technology industries such as semiconductors. Computers are considered high technology owing to their productive applications. See Hugh Patrick, "Japanese High Technology Industrial Policy in Comparative Context," in *Japan's High Technology Industries: Lessons and Limitation of Industrial Policy*, Hugh Patrick, ed. (Seattle: University of Washington, Press, 1986), 8.

10. For example, oil is a strategic commodity that is vital to virtually all sectors of economic activity. It is certainly vital to the survival of the modern economy as it is organized today.

11. John A. Young and others, *Global Competition: The New Reality*. The Report of the President's Commission on Industrial Competitiveness, Vol. II. (Washington, D.C.: Government Printing Office, 1985), 61.

12. Sherman Gee, *Technology Transfer, Innovation, and International Competitiveness* (New York: John Wiley, 1981), 5.

13. U.S. Department of Commerce, *Annual Survey of Manufacturers* (Washington,

D.C.: Government Printing Office, 1989), 43.

14. Scherer concluded that R&D has accounted for half the annual growth rate in productivity. Frederic M. Scherer, "Inter-Industry Technology Flows and Productivity Growth," *Review of Economics and Statistics* 64 (November 1982), 627-634. Also in Edward F. Denison, *The Sources of Economic Growth* (Washington D. C.: Committee for Economic Development, 1962) and Zvi Griliches, "Research Expenditures and Growth Accounting," in *Science and Technology in Economic Growth*, B. R. Williams, ed. (New York: John Wiley, 1973) the authors arrive at slightly lower estimates, but Denison comes closer to the productivity estimates of Scherer when he considers the inter industry spillovers and the asset features of R&D. Edward F. Denison, *Accounting For Slower Economic Growth: The United States in the 1970's* (Washington, D.C.: Brookings Institution, 1979), 122.

15. Organization for Economic Cooperation and Development, *Industrial Policy in OECD Countries* (Paris: OECD, 1990), 120; Richard Goodman and Julian Pavon, eds., *Planning for National Technology Policy* (New York: Praeger Special Studies, 1984); George Tolley, James Hodge, and James Oehmke, eds., *The Economics of R&D Policy* (New York: Praeger Special Studies, 1985); Motoshige Itoh and Kazuharu Kiyono, eds., *Economic Analysis of Industrial Policy* (New York: Academic Press, 1991).

16. Martin Baily and Alok Chakrabarti, *Innovation and the Productivity Crisis* (Washington, D.C.: Brookings Institution, 1988). See also U.S. Congressional Budget Office, *Federal Support for R&D and Innovation* (Washington, D.C.: Government Printing Office, 1984), 27ff; Keith Bradsher, "High Tech Industry is Hard to Help With Subsidy," *New York Times*, 2 February 1993, C1.

17. A condition of underinvestment in R&D may be exacerbated by the conflict generated by often daunting front-end R&D requirements and short-term business management preoccupation for returns measured by quarterly rather than annual earnings.

18. The inefficiency of the market provision for new technology has been established in many theoretical contexts. See Zvi Griliches, "Issues in Assessing the Contribution of R&D to Productivity Growth," *Bell Journal of Economics* 10 (1979), 92-116; Michael A. Spence, "Cost Reduction, Competition, and Industry Performance," *Econometrica* 52 (1984), 101-121.

19. Michael L. Dertouzos, Richard K. Lester, and Robert M. Solow, *Made in America: Regaining the Productivity Edge* (Cambridge: MIT Press, 1989), 33.

20. According to the same report, manufacturing industries were more productive and offered higher compensation than most service industries. U.S. Bureau of Labor Statistics, Department of Labor, *Employment and Earnings* (Washington, D.C.: Bureau of Labor Statistics, March 1991), 11.

21. U.S. Department of Commerce, *Annual Survey of Manufacturers* (Washington, D.C.: Government Printing Office, 1989), 9.

22. Lawrence Mishel and David Frankel, *The State of Working America* (Washington, D.C.: Economic Policy Institute, 1990), 94.

23. Stephen Cohen and John Zysman, *Manufacturing Matters: The Myth of the Post Industrial Economy* (New York: Basic Books, 1987), 102-106.

24. Organization for Economic Cooperation and Development, *Structural Competitiveness and National Systems of Production* (Paris: OECD, 1991), 147.

25. H. W. Coover, "Programmed Innovation Strategy for Success," in *The Positive Sum Strategy: Harnessing Technology for Economic Growth*, Nathan Rosenberg and Ralph Landau, eds. (Washington, D.C.: National Academy Press, 1986), 399.

26. U.S. International Trade Administration, *An Assessment of U.S. Competitiveness in*

High Technology Industries (Washington, D.C.: Government Printing Office, 1984), 4.

27. In 1988 the Pentagon published a study recommending the formation of institutions to meet the problem of competitiveness in the civilian sector. It included the formation of a Defense Manufacturing Board, a database on industrial developments, and a "production base advocate" in the Pentagon. See "Pentagon's Push to Bolster Competitiveness," *Challenges*, September 1988, 2. For more substantial data see Committee on the Role of the Manufacturing Technology Program in the Defense Industrial Base and Commission on Engineering and Technical Systems, *Manufacturing Technology: Cornerstone of a Renewed Defense Industrial Base* (Washington, D.C.: Office of the Secretary of Defense, 1987).

28. Trade relations between the United States and Europe have become increasingly colored by not so idle threats to levy tariffs on the other in the absence of accommodation. See James Sterngold, "A Frenzy over Spoils," *New York Times*, 25 October 1992, 10.

29. U.S. Office of Technology Asessment, *The Defense Technology Base: A Special Report* (Washington, D.C.: Government Printing Office, March 1988), 16.

30. Ibid., 15.

31. New Japanese chip equipment tends to be available to Japanese customers about six months or more ahead of overseas companies. U.S. General Accounting Office, *U.S. Business Access to Certain Foreign State-of-the-Art Technology* (Washington, D.C.: Government Printing Office, 1991), 44. See also U.S. Defense Science Board to the Department of Defense, *Report of the Defense Science Board Task Force on Defense Semiconductor Dependency* (Washington, D.C.: Office of the Under Secretary of Defense for Acquisition, February 1987). For general treatment of this issue see Shintaro Ishihara, *The Japan That Can Say No*, translated by Frank Baldwin (New York: Simon & Schuster, 1991).

32. A 1990 Harvard Business School study stated that DARPA's (Defense Advanced Research Projects Agency) contribution to computer development was amazing and almost single handedly created U.S. leadership in that industry, as quoted in John Tirman, ed., *The Militarization of High Technology* (Cambridge: Ballinger Publishing Company, 1984), 215. Kenneth Flamm argues that the government picking of the computer industry is a first-class success story. R&D and procurement by the government brought hardware costs down at an annual rate of 28%. In the 1940's and 1950's, most computer R&D came from five agencies—the Atomic Energy Commission, the National Science Foundation, NASA, the National Institutes for Health, and (clearly the most substantial) from the Department of Defense through DARPA and the Office of Naval Research. Flamm argues that only government could have provided the entrepreneurial intensity since it was a high-risk innovative process and guarding secrets was difficult. See Kenneth Flamm, *Targeting the Computer: Government Support and International Competition* (Washington, D.C.: Brookings Institution, 1987), 9, 18.

33. See James M. Utterback and Albert E. Murray, *The Influence of Defense Procurement and Sponsorship of Research and Development of the Civilian Electronics Industry* (Cambridge: MIT Center for Policy Alternatives, 1977), 3. See also Michael Borrus, *Competing for Control: America's Stake in Microelectronics* (Cambridge: Ballinger, 1988), 86-107.

34. Examples include gallium-arsenide components, massively parallel computing.

35. For example, 70% of the money the department of defense spent on semiconductor research in the mid-1980's went toward creating a nuclear resistant microchip. See Robert Kuttner, *The End of Laissez-Faire: National Purpose and the Global Economy After the Cold War* (New York: Alfred A. Knopf, 1991), 205.

36. John A. Young and others, *Global Competition: The New Reality*, The Report of the President's Commission on Industrial Competitiveness, Vol II. (Washington, D.C.: Government Printing Office, 1985), 186.

37. Since 1980, high-volume digital products have grown from 5 to 45% of Japanese electronics production, accounting for virtually all of the growth in domestic Japanese consumption of integrated circuits. Theodore H. Moran, "The Globalization of America's Defense Industries: Managing the Threat of Foreign Dependence," *International Security* 15 (Summer 1990), 57-100; Dataquest Incorporated and Quick, Finan, and Associates, *The Drive for Dominance: Strategic Options for Japan's Semiconductor Industry* (San Jose: Dataquest, 1988), 4-7.

38. The common perception that the sophisticated electronic systems that underpin American military performance employ the most advanced semiconductor devices conflicts with the fact that commercial electronic sectors have become the primary drivers of microchip innovation and within which the most sophisticated devices are often used. See John W. Kanz, "Technology, Globalization, and Defense: Military Electronic Strategies in a Changing World," *International Journal of Technology Management* 8 (1993), 59-76.

39. U.S. International Trade Commission, *Foreign Industrial Targeting and Its Effects on U.S. Industries, Phase I: Japan*, USITC Publication no. 1437 (Washington, D.C.: Government Printing Office, 1983); U.S. International Trade Commission, *Foreign Industrial Targeting and Its Effects on U.S. Industries, Phase II: Europe*, USITC Publication no. 1437 (Washington, D.C.: Government Printing Office, 1983).

40. R&D expenditures are a sunk cost. The more goods the company can produce, the cheaper each successive product becomes because most of the cost is the front-end investment in R&D. See Luc Soete, "Technological Change and International Trade," in *Technical Change and Economic Theory*, Giovanni Dosi and others, eds. (New York: Pinter Publishers, 1988), 421.

41. Smaller firms have a history of being more innovative and flexible than their larger counterparts. However, smaller enterprises lack the production economies of larger organizations and are more vulnerable to the vicissitudes of short product life cycles and cyclical demand. For example, in the semiconductor industry, the cost of developing a device can be more than $75 million, and to construct a state-of-the-art manufacturing facility can run more than $800 million.

42. Throughout modern history countries have tied their power and prosperity to the development of strategic sectors. In the nineteenth and early twentieth century coal and steel were symbols of national power.

43. Stephen Cohen and John Zysman, *Manufacturing Matters: The Myth of the Post-Industrial Economy* (New York: Basic Books, 1987), 239.

44. U.S. National Advisory Committee on Semiconductors, *A Strategic Industry at Risk* (Washington, D.C.: Government Printing Office, 1989), 5.

45. U.S. Defense Science Board to the Department of Defense, *Report of the Defense Science Board Task Force on Defense Semiconductor Dependency* (Washington, D.C.: Office of the Under Secretary of Defense for Acquisition, February 1987), 43.

46. U.S. National Research Council, *U.S.-Japan Strategic Alliances in the Semiconductor Industry: Technology Transfer, Competition, and Public Policy* (Washington, D.C.: National Academy Press, 1992), 85.

47. U.S. National Research Council, *Race for the New Frontier: International Competition in Advanced Technology* (New York: Simon & Schuster, 1984), 6.

3

The Industrial Policy Debate

The declining performance of American firms in high technology product markets and a general sense systemic economic lethargy have rekindled the debate over the most efficient or least desirable roles the government should assume in promoting a healthy economy. It is a debate that has become richly infused with a varied historical treatment of industrial policy, the weight of political ideology, pork-barrel politics, and empirical and theoretical economic analyses that examine the wisdom of government industrial targeting. Industrial policy advocates espouse a strategic orientation to government intervention. These proponents argue that the United States cannot afford to remain passive while governments in Japan and Europe continue to change competitive conditions at the expense of American industry.[1] The suspicion that foreign industrial policies have conferred unfair advantage has led to a rising sentiment that such programs should be countered by adoption of similar American industrial policies. Those who favor a laissez-faire orientation regard this contention as specious and an oversimplification of the evidence. There are many political and economic problems that inhere to government promotion and there are numerous examples of industrial policy failing or even being counterproductive. There is reason to believe, therefore, that many claims of foreign industrial policy success are exaggerated, and the contention that government can make decisions more effectively than the market remains a questionable one. Preferring the clarity of the market place, industrial policy opponents stress that the choice is fundamentally between an imperfect market[2] and a more imperfect government.

WHAT IS INDUSTRIAL POLICY?

Every modern economy is to a greater or lesser degree influenced by government regulations, tax policies, promotional subsidies, procurement policies, and antitrust laws. In addition, macroeconomic policies that increase the quality and quantity of factors of production, such as labor, capital, natural resources, and technology (i.e., incentives to save, invest, public education, R&D), affect the ability to compete in international markets. Some policies are explicit in their goals while others have indirect and sometimes unanticipated effects on the economic and industrial structure. Government has had and will continue to have a major impact on economic activity. Owing to the already pervasive influence of government on economic activity, there is some definitional confusion as to what specifically is industrial policy, as the term has gained so many different meanings over time that "today it has lost nearly all meaning."[3] It is evident that industrial policy can assume various forms and mean different things to different people, representing any and all governmental intrusions into the economy, or it may embrace more specific objectives such as the correction of perceived market failures and the promotion of a particular economic activity. The objective of government policy may be to reallocate resources out of some existing activity to enhance efficiency, to maintain current employment and or price levels in a sector contrary to the impulse of the market, to facilitate a change in the composition of national industrial production.

A general allusion to government intervention in the economy does not capture the underlying spirit of the international debate on industrial policy, and it does not address the goals that make an industry important to a nation. Every government endeavors to influence macroeconomic fundamentals in an optimal fashion to promote the business and economic interests of the nation. Failure to do so is an abdication of responsible government. This process is different from an explicit industrial policy, one that identifies sectors of the economy that warrant special treatment—ones that benefit from government promotion with access to resources that the market might otherwise not provide. Providing government subsidy or trade protection to economic agents that are presumed to be competing in the free market cuts against the official doctrine of free and unfettered trade fashioned at Bretton Woods after World War II, and even if it is welfare reducing in the aggregate, it can alter the distribution of spoils in a particular industry and for this reason has been typically associated with unfair trade practices. The U.S. International Trade Commission (USITC) defines *industrial policy* as "coordinated government actions that direct productive resources to give domestic producers in selected industries a competitive advantage."[4] This requires that productive resources are directed (hence not only trade protection), that only selected industries be directly benefited, and that the purpose of targeting be to provide domestic producers an advantage in the selected industries. As such, industrial policy is government intervention directed toward a micro level objective, a policy distinguished by the differential treatment that a sector or industry receives, and this orientation will guide our examination of industrial policies designed to compete in high technology markets.

An industrial policy should, presumably, nurture and protect upcoming industries or regulate and assist the shift of resources away from declining industries in order to maximize economic growth. Although government targeting typically involves high technology industries, political realities are such that protection for suboptimal sectors will probably remain.[5] The logic of a national industrial policy has been questioned in an era of ever greater globalization. An important issue to consider is the extent to which one can capture the putative benefits of promotion. For our purposes, an industrial policy as a public policy tool has utility only if it is possible to contain benefits, to a greater or lesser extent, within national borders.

The Globalization Variable

The trend toward internationalization of research and production activities raises questions about the ability of national promotion to secure a domestic advantage in particular industries. The same forces responsible for multinational corporations, strategic alliances, and joint ventures complicate the presumption of assuming a national technology position. The concept of national competitiveness is more complex now than ever before. Manufactured products are created from factor inputs sourced all over the world, and the extent to which we are able to print a national stamp on such a process has, therefore, become increasingly difficult. The former chairman of the National Semiconductor Corporation offered "we are using Russian engineers living in Israel to design chips that are made in America and then assembled in Asia."[6] This begs the question; where do we draw the domestic label and, by extension, does the blurring of the national origin of goods compromise the integrity of a national agenda? The decreased correlation between a producer's home and its value-added has contributed to the blurred distinction of national origin. For example, domestic subsidiaries of foreign firms operating in America account for close to 8% of American manufactured goods exported; even aerospace, an industry hotly contested by national parties, is not immune from the forces of internationalization. The American content of Boeing aircraft is approximately 70% if Rolls-Royce engines are employed, and some versions of the Airbus represent factor inputs that are as much American as European.[7] Innovations today are no longer isolated by geography; they are the product of technology fusion dependent upon the blending of technologies from different industrial R&D efforts. Innovation and technological sophistication are very real assets to a national economy, but the process is not easily confined to a national network.[8]

Consequently, the wisdom of a technology industrial policy is immediately suspect because many companies can, by virtue of their multinationalization, simply shift assets overseas, thereby diminishing the putative benefits of industrial promotion at great cost to the American taxpayer.[9] The transnational character of most industries suggests that national differences may be less relevant in an increasingly borderless global economy.[10] However, the interlinked world economy is still not yet a reality, and the contention that the international corporation has no country and that the nation state is "just about through as an economic unit" is decidedly premature.[11] Borders still matter, since companies remain dominated by a headquarters mentality, markets are protected, and not all assets and jobs are transferable.

The firm has not transcended the nation. Country-specific variables will still have considerable impact on the nature of the inputs sourced there. Factors such as natural resources, infrastructure, education, and technical expertise will, to varying degrees, influence the extent to which a nation participates in the wealth-making process of the global economy. The economic reward for any economy depends on the nature and quality of the inputs going into global products. Competitive advantage is created and sustained through differences in national economic structures, values, cultures, and institutions. Indeed one notable economist who places a premium on defining competition at the firm level concedes that "the home nation takes on growing significance because it is the source of skills and technology that underpin competitive advantage . . . and stimulates the greatest positive influences on other linked domestic industries."[12] First, what the global medium lacks is the practice and know-how that is spread less formally in a social milieu. Second, the aggregation of high technology industries in certain areas such as the SiliconValley, Route 128 near Boston, and Aerospace in Southern California produce synergies and facilitate the exchange of ideas and trade in goods. Third, a common nationality benefits local producers and technologies through such procedures as state-wide "Buy American" campaigns and the government procurement process. Cultural mores, technical standards, and language continue to form relationships colored by proximity.

The process of technological diffusion provides the local economy with an early edge. Domestic suppliers and producers benefit earlier and more rapidly from domestic innovations than do foreign users. This remains the case largely because domestic firms are in a better position to capture that innovation. They typically are more integrated into a national framework, have contacts, read trade journals in their own language, and have the advantage of being at the center of the diffusion process. A large lead time is not necessary. A producer of semiconductor manufacturing equipment providing a merchant semiconductor manufacturer a several-months lead down its learning curve can make that semiconductor merchant a formidable player in the world market.

The customer-supplier-producer relationship is another important linkage that advanced transportation and communication technology has not yet completely severed. Proximity to the manufacturer offers suppliers and services a direct linkage in that it facilitates a faster turnaround time, the resolution of problems with alacrity, and typically better service. For a firm to be successful in the semiconductor industry it must maintain a close association with its customers, and geography and language remain central to developing specialized memory chips for computer customers. Users developing new systems and applications account for approximately 60% of the new ideas generated in the semiconductor industry, and the chip maker must be aware of the specifications required by the customer and typically work in concert with that client.[13]

American semiconductor manufacturers have been buoyed by the presence of a large domestic computer industry, and similarly, the Japanese semiconductor industry has been pushed along by the user demands of its large consumer electronics industry.[14] The importance of patronage cannot be overstated. It encourages innovation by guaranteeing a ready market, thereby reducing systemic risk, while also providing an opportunity to realize economies of scale. The American semiconductor and aerospace industry would have assumed a different course had the Department of Defense not provided a guaranteed market for its product in the 1950's and 1960's. Similarly, an important piece of Japanese semiconductor promotion was the closed-market procurement policies of Nippon Telephone and Telegraph, NTT.

The multinationalization of an industry does not necessarily mean that it is, nor cannot be, an instrument of national strategy. The manner in which technology is treated suggests that many such firms continue to have a strong home orientation. Japanese multinationals recognize that keeping the best technology and technical tasks at home gives them the advantage in meeting the next generation of technical challenges. Japanese technology does not typically emigrate until it is in a declining sector or has become more standardized, in the spirit of the Vernon Product Life Cycle. Even then, as in the case of Japanese Victor Corporation, which makes 60% of the VCRs that it sells in Europe, the critical high-tech component comes from Japan.[15] Similarly, a 1983 survey of 23 German multinational corporations showed that 83% of their R&D personnel was concentrated in the home nation, which otherwise accounted for only 65% of the total employment.[16] Du Pont has 90% of its R&D personnel in the United States compared with 65% of its total assets and 76% of its total employment.[17] The home nation remains the center of the multinational firm's innovative efforts, and it is where strategic and integrated decisions are made.

Japanese multinationals are not as bound to MITI as once before but are still linked back home through research facilities, consortia, financing, and suppliers. Even those firms completely independent of MITI, such as Sony, Matushita, and Kyocera, concentrate their research efforts at home. Both Sony and Matushita keep 90 and 99% of their engineers respectively at home, with their American facilities assuming a role of production support.[18] Although the sourcing of products has blurred the distinction of national origin, there remains a primary interest in what inputs one provides. The ability to attract the technical component gives the value-added advantage and is increasingly guarded most jealously.[19] The global economy is not yet and may never be completely international. The distinctions based upon the national context in which various enterprises operate and owe allegiance persist.

RHETORIC VS. REALITY: AMERICAN INDUSTRIAL POLICY

Government intervention in the American economy has been enormous despite an official adherence to the spirit of laissez-faire. Through tax and trade policies, credit programs, R&D programs, and state and local industrial policies, the American government has been and remains actively involved in influencing economic development. Promoting technologies and products for the military has been, perhaps, the single largest and most clearly defined role the government has played in sectoral economic development. The overriding factor was the emergence of the Cold War and a commitment to military preparedness. Accordingly, the federal share of total American R&D spending ballooned after World War II.[20] Coincident with the stated policy of containment was a scientific and technology strategy premised on maintaining superiority over the Soviet Union. This was reflected in the defense proportion of total federal R&D spending as Cold War military realities compelled the government to nurture those industries deemed critical to national defense.[21]

Procurement shapes an industry's product. With an annual budget of 5-7% of GDP in the 1980's, and more recently 3%-4% of GDP in the 1990's, defense procurement and technology development have acted as a defacto industrial policy—one biased toward military goods and systems. This breed of government promotion emphasizes performance over price and does little to enhance the competitive position of the commercial side of the industry. Whether it be energy or defense, the American government has intervened in economic sectors deemed critical to national security objectives.

There is also evidence that the American government regularly intervenes in the marketplace without any such national objective. The result is that the government spends billions of dollars, directly and indirectly, intervening in economic activity with policies that lack any central rationale or objective and that typically reflect the lobbying efforts of interested groups and industries.[22] One congressional committee stated that "this country has an extensive and expensive array of industrial policies. . . . Neither coordinated, cohesive, nor consistent . . . a melange of stop-gap measures."[23] The Urban Institute reported that there were 329 federal financial subsidy programs to businesses of which 81% were sectorally targeted.[24] The study discovered that these programs lacked rigorous review, were costly ($303.7 billion in 1980), neither cost nor task effective, and did not reflect any comprehensive objective.

The official doctrine of American trade policy is free trade, and there has been little assistance for export promotion. However, there has been increasing import restriction to protect, in many cases, politically influential but declining industries. Through the 1980's Republican presidents professed adherence to free-market principles yet tightened sugar import quotas, negotiated (some would say dictated) Voluntary Export Restraints (VERs) with the Japanese in the early 1980's for automobiles, and expanded relief against textile imports. In 1987 Treasury Secretary James Baker boasted that the Reagan administration had granted more import relief to U.S. industry than any of his predecessors in the previous half-century.[25] This is not laissez-faire economics. The United States has demonstrated a willingness to interfere in the market to provide relief for a specific industry or sector, but the government has not assumed a long-term perspective nor the responsibility to promote a vital economic structure in those industries.

Government intervention is also practiced with increasing regularity on the state and local levels. The Office of Technology Assessment concluded that "National industrial policy is already being implemented by the states . . . which have their own industrial policy."[26] In the last ten years state governments have gone beyond the simple task of luring jobs to their locale to the goal of creating jobs in their region. State and industry have begun working closely together in most of the 50 states. The federal government is also involved in state industrial policy. State and local governments spend billions of federal dollars annually on economic development programs, infrastructure development tied to industrial targeting, and tax expenditures. The federal government's outlay for associated state industrial policy activities increased 500% from 1975 to 1983, prompting a congressional review of spending patterns. Despite revelations that the federal government had financed questionable ventures, little was done to reform the system. As Otis Graham relates, "Washington was a partner to state industrial policies, but had no goals, and looked the other way."[27]

Given the extent of government intervention in the economy, the debate in the United States over whether or not we should have an industrial policy is potentially confusing. This record of government sectoral intervention contrasts with the low view former President Bush and previous American presidents have held of industrial policy.[28] For example, the Bush administration opposed the creation of a Department of Industry and Technology (the civilian equivalent of DARPA) and rejected support for high definition television.[29] Government promotion of technologies and products that serve a military mission has had a purpose and clarity that typically escapes equivalent civilian policy making. This is because the governing establishment has remained uncertain as to the merit and legitimacy of government targeting to promote competitiveness in selected industries without specific military priorities.

Whether or not this uncertainty is or has been detrimental to America's economic interests is subject to debate. The HDTV policy experience, for example, illustrated the absence of government institutions capable of dealing with complex technological issues that have both military and commercial applications, but it also exemplified the merits of caution. The sales of HDTV, projected to be in excess of $170 billion by the year 2010, may eventually be one of the largest sources of demand in the electronic food chain. Governments in Japan and Europe have heavily subsidized HDTV research and development. A European industry and government consortium has committed over $200 million to the effort. The Japanese presented a prototype system as early as 1986, thanks to a $500 million consortium created in 1981. In 1986 the Federal Communications Commission endorsed the Japanese standard, thereby consigning control of HDTV design, technology, and an enormous new market to the Japanese. European outrage encouraged the FCC to reverse its position.[30] The American effort to capture a potentially important market has remained comparatively small owing to the reluctance of the government to intervene and the existence of only one American television manufacturer, Zenith.

The HDTV debate demonstrates the complexity and hazards of government promotion of technological competitiveness. Industrial policy proponents argued that because HDTV was a technology driver for the entire electronics industry, special assistance from government was necessary to create an American HDTV industry to preserve the competitiveness of related industries. However, the importance of HDTV may have been exaggerated.[31] The contention that HDTV is the principal path for advancing sophisticated digital technologies and future information systems is questionable. It may not even be the best one. In addition, the United States benefited by its inability to organize a national drive toward HDTV in the mid-1980's. The Europeans and Japanese committed themselves to what now appears to be an inferior analog standard while the unforeseen advances of American research have enabled the United States to leap-frog their efforts by adopting a more sophisticated digital standard. European and Japanese governments now confront the possibility that their government outlays to develop HDTV analog components may be wasted. The HDTV experience illustrates that even in cases where industrial policy proponents are most vociferous, it is not certain that the reallocation of private and public resources is either necessary or desirable.

PERSPECTIVES ON INDUSTRIAL POLICY

The idea of institutionalizing an industrial policy has precipitated an ideological debate in the United States not experienced in Japan or Europe. Opponents argue that government intervention is distortive, that it cannot improve on the performance of the marketplace, and that raising the necessary funds for industrial promotion imposes a further cost on the economy.[32] These potential costs and inefficiencies of an industrial policy are compounded when one considers the penchant for pork barrel in American politics. Although a more sensible governance of the promotion apparatus already in place is arguably warranted, the industrial policy debate in the United States is not specifically about replacing the hodgepodge of existing policies with coherent and targeted strategies. Rather, it involves the proper role of government in promoting competitiveness in economic sectors of particular importance to economic power. Since this conflicts with the ideological and institutional framework of American economic policy, the promise of industrial policy must move beyond claims of theoretical advantage and sustain a more rigorous analysis of practical applicability.

Implementation

A principal challenge for an industrial policy agenda is to enact policy without succumbing to pork-barrel politics. To the extent that this may be the case, the issue of political economy becomes a domestic concern that warrants caution on behalf of an active industrial policy. The question is how one addresses the problem of interest-group influence on decision making by policy makers. How would inevitable mistakes be corrected in a political process that is even more averse to admitting failure than the decision-making process of large corporations? Both Spar and Vernon, in *Beyond Globalism: Remaking American Foreign Economic Policy*, and Krueger, in "Theory and Practice of Commercial Policy: 1945-1990," make the case that America is particularly vulnerable to special interests and pork-barrel politics.[33] In *The Case for Industrial Policies*, even Lester Thurow admits that the politicization of industrial policy by interest groups and politicians eager for funds for their district could render any industrial policy a wasteful and futile exercise.[34]

The challenge of efficient policy implementation is an important issue in the industrial policy debate. Badaracco and Yoffie argue in "Industrial Policy: It Can't Happen Here" that regardless of the merits of a policy on paper, political realities in the United States will frustrate any attempt to implement an industrial policy.[35] They contend that industrial policies in Japan have had some success owing to unique political and economic conditions. Industrial policies in Europe, on the other hand, have been implemented in a similar economic and political climate as the United States and the record of success there is far less encouraging. In *Japan as Number 1: Lessons for America*, Ezra Vogel suggests that political manipulation could be diminished if the government created an independent political body of highly paid (to encourage them to stay) senior-level bureaucrats with wide freedom to implement long-term industrial programs. Vogel concedes that an effective industrial policy is impossible without a cadre of de-politicized professional bureaucrats. However, it is questionable that such a cadre can ever exist. Even the Japanese economic and political system, long synonymous with stability and sound economic policy, created systemic corruption that recently contributed, if not precipitated, the first-ever fall from power of the Liberal Democratic Party.

The challenge of policy implementation is generally conceded by most industrial policy advocates. Bruce Scott commented in "National Strategies: Key to Economic Competition" that the development of a coherent industrial policy in America is a political rather than an analytical challenge.[36] Robert Reich, a leading spokesman for industrial policy, explains in the *Work of Nations* that industrial policy is not national planning *per se*, but policies to make the economy more adaptable and dynamic. Yet whether the rubric is infrastructure, human capital, or industrial development, the impulse for private political interests to secure private gain at the expense of sound economic judgment could make matters worse. In "Why the U.S. Needs an Industrial Policy" Reich argues that industrial policymakers can remain above the fray of political interest by securing broad public consensus for their programs. Robert Reich does not adequately articulate how consensus and inclusion will diminish this problem of politicization. Consensus building itself is a political process, and as such it is impossible to stay above the fray. In "Industrial Policy: A Solution in Search of a Problem," Charles Schultze questioned the ability of government to pick winners and acquiesce in the declining fortunes of other sectors. Since government is "lacking defensible criteria for overriding political pressures when protecting losers, the government chooses poorly."[37] In *Minding America's Business*, Robert Reich and Ira Magaziner indicate that our standard of living can rise only if capital and labor flow increasingly to industries with high value-added per worker. Yet the criteria of value added per worker is potentially misleading. The cigarette and petroleum industries have higher value-added per worker than aerospace, but it is clear that the authors favored a high technology industrial policy and not one that would channel funds into Marlboro or Exxon.

In *The Technology Pork Barrel*, Cohen and Noll doubt the necessity of strategic promotion and outline previous efforts in which government promotion of commercial technology have been economic failures.[38] Failures such as the supersonic transport, the Clinch Breeder reactor, the space shuttle, and the synthetic fuels program suggest in their view that even a limited strategic industries program would have limited effectiveness. The authors question the ability of the government to select critical technologies on economic merit independent of political influence, and therefore conclude that government promotion may not only promote the wrong technologies but divert resources away from other promising technologies. Laura D'Andrea Tyson submits in *Who's Bashing Whom: Trade Conflict in High Technology Industries* that political and economic risks may be minimized with appropriate institutions and incentives.[39] Tyson argues for industry-led projects and government funds to be released for promising areas only after, agreeing with Vogel, they are reviewed by a politically independent panel. She points out that the Advanced Technology Program (ATP) in the Commerce Department and the program for Cooperative Research and Development Agreements (CRADA) between national laboratories and the private sector already meet the above criteria.[40]

The feasible implementation of industrial policy in America thus remains a contentious issue. Theodore Eismeier represents the view of many opponents in the "Case Against Industrial Policy" when he suggests such a policy would result in the explosion of ill-conceived government programs and waste. George Lodge counters in *Perestroika for America* that industrial policy is not about big government.[41] It is about promoting competitiveness. He continues that economic and political abuse may be contained and government largesse prevented through cost-sharing. Placing private money at risk is a sound method of promoting such responsibility. Ensuring efficient and effective implementation will probably remain an operational necessity before any industrial policy program can be accepted in the United States.

Macroeconomics Considered

Macroeconomists argue that the competitiveness issue begins and ends with the economic environment in which firms operate. Most presume that markets are sufficiently competitive and that firms strive to be as efficient as possible. Accordingly, the Japanese economic miracle is, therefore, not the result of well-conceived industrial policies but a high investment rate, low capital costs compared to the United States, a better trained and educated workforce, and other favorable conditions that involve the corporate and social cultures. Firms in the United States have been losing out to foreign concerns precisely because of the high cost of capital, the enormous budget deficit, and an inferior education that affects productivity. These macro conditions encourage a short-term orientation to return on investment and a lower rate of investment in plant and equipment. By extension, the competitiveness of strategic industries cannot be guaranteed through government manipulation, but by improving those macroeconomic conditions on which success depends.

Herbert Stein, the former chairman of the Council of Economic Advisors, questioned the importance of manipulating industrial structure at all. He stated that "if the most efficient way for the U.S. to get steel is to produce tapes of "Dallas" and sell them to the Japanese, then producing tapes of "Dallas" is our basic industry."[42] In "Can America Compete?" Robert Z. Lawrence examined the structural changes in American industry and concluded that the United States was not de-industrializing as industrial policy proponents had suggested. Lawrence admitted that the share of manufacturing employment from 1950 to 1982 had dipped from 35 to 21% but because manufacturing share of output had remained basically unchanged, 25 to 23.5% of total output, he concluded that fears of an erosion of the American manufacturing base were unfounded.[43] Lawrence stated that economic performance is tied to macroeconomic conditions and regarded industrial policy efforts as ones to protect and subsidize firms for fear that they cannot compete.

Critics of an American industrial policy have reaffirmed their belief in market principles, argued that intervention was premised on faulty economic theory, and articulated conflicting lessons from industrial policies in America and elsewhere. In their view, industrial policy does not work because only the market is capable of determining the appropriate economic structure. By extension, government intervention is distortive, welfare reducing, and it is incapable of making efficient decisions concerning the allocation of resources. Government should not, therefore, interfere with the market mechanism.

Lessons on Industrial Policy

Central to the position of industrial policy advocates is the perception that free-market forces are creating an economic structure in the United States that is inferior to those engineered by governments abroad. Lester Thurow concluded in *Head to Head: The Coming Economic Battle Among Japan, Europe and America* that the United States risks losing in this competition because governments in Japan and Europe are creating a competitive infrastructure and promoting high technology value-added industries on which wealth and prosperity have been linked. Although Thurow overstates the success of government engineering of economic prosperity, this view that the United States is being outsmarted by foreign governments is a *cause célèbre* for industrial policy advocates. By extension, the purported success of foreign industrial policies provides a case for similar policies to promote American competitiveness. Lessons from home and abroad have thus added to the industrial policy debate.

A central issue for industrial policy advocates is that the government should augment the system of capital formation for certain industries in order to promote competitiveness. Some attention has focused, therefore, on the record of the Reconstruction Finance Corporation (RFC). The RFC was a government development bank created by President Roosevelt during the Depression and existed until 1953. The agency disbursed $40.5 billion, and proponents of industrial policy insist that it shored up the banking system during the Depression, extended much-needed capital to railroads and businesses, and financed public works.[44] Opponents of industrial policy view it quite differently. Murray Widenbaum and Michael Athey stated in "What is the Rust Belt's Problem" that "the RFC shows that government subsidy of business encourages a misallocation of resources and provides opportunities for political favoritism."[45] Melvyn Krauss submitted in "Europeanizing the U.S. Economy: The Enduring Appeal of the Corporatist State" that the RFC was "inspired by Mussolini's Instituto per la Riconstruzione Industriale(IRI). . . . European experience shows that the real myth is the notion of an efficient industrial policy in the first place."[46]

Those who view the RFC as a grand example of government's assisting industries to reindustrialize and boost productivity and those who view it as a failure are both correct. The RFC progressed through three distinct periods over the course of two decades. In the first period, during the Depression, it carried out important activities in providing liquidity to the banking system and capital to production facilities. In the second period, during World War II, it facilitated the formation of industries critical to the war effort. The third phase, after the war, with a robust economy and the ascendance of foreign policy issues, the RFC wandered for lack of a coherent objective.[47] Without a clearly defined agenda the RFC was terminated by Congress in 1954.

The progress of industrial policy in the United Kingdom was viewed with particular interest by American scholars of industrial policy, given the similarities of the two nations.[48] In "Britain's Economic Performance," Richard Caves and Lawrence Krause argue that efforts to reverse the "British Disease" through the implementation of industrial policies has largely failed.[49] In large measure, the British industrial plan failed because, despite intentions to the contrary, the government spent disproportionate sums propping up sunset industries and bailing out losers.[50] A similar conclusion was reached by Raymond Vernon in "Enterprise and Government in Western Europe."[51]

European industrial policy failures are not confined to Britain.[52] Efforts to promote "national champions" in France have been very costly and ineffective. According to Jean-Claude Derian,[53] this policy has not challenged American and Japanese dominance because it sheltered industries from foreign and domestic competition and therefore did not promote necessary efficiencies. A former EC Commissioner, Etienne Davignon, commented that in "America high technology makes money and in Europe it costs money."[54] Yet Theodore Eismeier, an opponent of industrial policy, called for a balanced treatment of the foreign record. He states the "broad diversity of foreign experience permits no easy generalization. . . . The evidence is decidedly mixed and ought not to be overstated."[55]

The enviable record of economic success in Japan is often cited as evidence of successful promotion of strategic industries. That Japan experienced phenomenal postwar economic growth is not disputed. What is contended by the industrial policy opponents is the purported cause for such economic growth. Philip Trezise argued that to "attribute to industrial policy a crucial role is an expression of faith, not an argument supported by discernible facts."[56] He states that public funds in Japan have not been directed in any sizable requirements to economic sectors with high growth potential. The Ministry of International Trade and Industry (MITI) had a small role in the Japanese economic miracle and did not preside over an elaborate process of picking winners and supporting strategic industries.

Takatoshi Itoh states in *The Japanese Economy* that inadequate attention is given to the fact that MITI made mistakes and that many industries became successful without government assistance, and in some cases despite government.[57] MITI has attempted to limit firm entry in a number of important industries because of a concern with achieving economies of scale and avoiding excessive domestic rivalry. In its attempt to cartelize the automobile industry in the 1960's, for example, firms such as Honda persevered despite pressure by MITI to stay out of the industry and has since become one of the world's largest automakers.[58] MITI has had success in many high technology industries by stimulating demand, encouraging domestic rivalry, providing technical and monetary assistance, and organizing cooperative research projects. However, MITI has failed in other areas where it has attempted to micromanage the industrial structure by encouraging mergers, sanctioning cartels, and maintaining inefficiencies by insulating some industries from international competition.[59] On balance, Japanese industrial policy has succeeded only in sectors with effective domestic rivalry, which in some cases has substituted for international competition.

Scholars emphasize to a greater or lesser extent the role of industrial policy in the Japanese economic miracle. In *MITI and the Japanese Miracle: The Growth of Industrial Policy 1925-1975*, Chalmers Johnson places great emphasis on the ability of the Japanese government to guide large resources into strategic sectors. In *Shadows of the Rising Sun: A Critical View of the Japanese Miracle*, Jared Taylor argued that industrial policy has worked in Japan but would not be appropriate for the United States because it exacts a social cost amenable only to Japanese values and institutions.[60] Robert Ozaki insisted in "How the Japanese Industrial Policy Works" that industrial policy is part of a national economic and social fabric and that the United States lacks the appropriate framework to make those lessons transferable.[61]

The distinct milieu in which Japanese industrial policy was implemented is an important consideration. Few observers would argue that the Japanese government has not had an industrial policy, but many doubt the central importance of specific policies. More emphasis is placed on particular strengths of the private sector and sound macroeconomic policies.[62] Accordingly, a low cost of capital, long-term vision, social solidarity, and aggressive corporate behavior[63] can be considered the principal sources of Japanese success. In *Between MITI and the Market: Japanese Industrial Policy for High Technology*, Daniel Okimoto attributes the effective Japanese industrial policy in high technology to the overall system in which it operates and that no single policy or factor can be viewed as decisive.[64] It is the combination of distinctive factors[65] that has made the Japanese industrial policy an apparent success. Okimoto concedes that the weight of evidence suggests that MITI facilitated the development of Japan's high technology industries, but he submits that there has not been a clear scientific method to establish the causal connection between industrial policy and competitiveness in high technology. Okimoto concludes that "while it is hard to demonstrate beyond a reasonable doubt that the instruments of industrial policy have given Japanese companies their competitive edge . . . it has functioned as an indispensable mechanism in the political economy of Japan."[66]

In "Japanese High Technology Industrial Policy in Comparative Context," Hugh Patrick argues that Japanese industrial policy has been moderately useful but has not been the central reason for economic success.[67] Patrick places more emphasis on tax incentives to engage in R&D and invest in emerging industries, incentives to save and a low cost of capital, and a highly effective public education system. Patrick maintains that Japan does have a high technology industrial policy, but the resources allocated are modest and their effectiveness not clearly established. Even conceding that industrial policy was partly responsible for the economic miracle, Patrick continues that the policy of picking winners was a process of catching up suited only to a nation in a follower position. MITI has demonstrated less influence in industrial matters since Japan has effectively caught up and its industries have become world-class producers. Ken-ichi Imai states in "Japan's Industrial Policy for High Technology Industry" that Japanese business is responsive to non-price signals, particularly from the government or other consensus-forming organizations. He states that MITI continues to exercise industrial policy around guidance and what he terms soft policies.[68]

Lessons offered by the Japanese experience have been interpreted differently. Vogel and Johnson believe that with several modifications, a Japanese type policy may be effectively used in the United States. Ozaki and Taylor argue that the Japanese system of subsidies, protection, and strategic guidance has worked in Japan but cannot work in the United States. Patrick and Okimoto agree on the issue of transferability but question the central importance of industrial policy in the Japanese model. Richard Nelson and George Eads suggest in "Japanese High Technology Policy: What Lessons for the U.S.?" that the most important lesson Japan offers is that any nation that aspires to strength in high technology industries should promote technical education and macroeconomic policies that support economic growth more generally.[69] Nelson and Eads emphasize the importance of macroeconomic fundamentals, but they also conclude that the Japanese experience "confirms the value of government support for generic technology, and given the declining spin-offs of military technology there are strong reasons to establish a basis of support in America, one independent of the Department of Defense (DOD)."[70] Hugh Patrick concurs. He states that the "risks, costs, and inability to appropriate fully the benefits of R&D mean government funding of R&D can be desirable, in both Japan and the United States."[71] This view is not supported by Stein, Lawrence, and Schultze, who consider any industrial policy as inherently inferior to the efficiencies of the free market.

Consistent with this is the notion that government cannot create competitive industries. In *The Competitive Advantage of Nations*, Michael Porter considers economic competition as a company-level concern. He argues that nations do not compete, companies and industries compete and as such the role of government is necessarily limited. He acknowledges that government has a role in shaping the quality of the inputs that firms draw upon and the institutional structure in which they operate, but maintains that subsidy, protection without domestic rivalry, and government procurement will not foster productive national firms. Porter concedes, however, that industries and technologies that affect the potential productivity of many other industries deserve special attention.[72]

The preceding illustrative survey of the major issues in the industrial policy debate suggests that there is no broad agreement that industrial policy is the answer to America's economic problems. Many argue for more government-industry cooperation, but there is little support for an MITI-type mandate or a adoption of the industrial policies practiced by European states.[73] Although the role of MITI has diminished, it still assumes a major role in coordinating and promoting research in next-generation technologies and, according to Christopher Freeman in *Technology Policy and Economic Performance*, this explains their continued success in various high technology fields.[74] What re-emerges is the more general question of whether the American government needs a mechanism to coordinate and support R&D in emerging technologies and in areas where the United States is behind.

There is general acceptance that market failures exist in R&D and related activities, but far less acceptance as to whether government programs can be managed efficiently. Cohen and Noll carry this point a step further. They concede that conditions of underinvestment and market imperfections may exist, but conclude that there is no simple method to determine underinvestment in certain commercial technologies and lacking a scientific method in the selection process increases the likelihood that the government will choose poorly. Even assuming proper selection and appropriate funding, there are those who believe that technonationalism is an inappropriate orientation in the late twentieth century. However, the national technological infrastructure is an important source of competitive strength, and an examination of the globalization variable indicates that linkages associated with geographic proximity persist. Technology accrues locally before diffusing internationally. Technological know-how accumulates in the production networks of suppliers and contractors, in social networks among technical peers, and in the cumulative skills of the workforce. This view is supported by Paul Krugman in *The Age of Diminished Expectations: U.S. Economic Policy in the 1990's*, in which he advocates an active though cautious approach to industrial policy. Optimal national competitiveness involves a mix of institutions, some private and some public. After all, the ability of government to provide for an efficient infrastructure and education has always been an important component of economic progress. However, industrial policy advocates conclude that the public sector should assume a greater role, not only with regard to infrastructure but also in the promotion of certain important commercial technologies. The new trade theory places government-industry collaboration in a theoretical framework that challenges the notion of free market optimal efficiency. It recognizes that if there are large up-front R&D costs and scale advantage, then government policy may make an enormous difference in determining which nation ultimately dominates an industry.

NOTES

1. Through its financing of large ventures such as Ariane (space), Eureka (electronics), Jessi (semiconductors), and ESPRIT (information technology), and a long-standing commitment to Airbus, Europe is preparing itself to meet the competitive challenge of the future. Likewise, the Japanese Ministry of International Trade and Industry (MITI) has had a tremendous influence on the composition of Japanese industry. It acts as a catalyst in identifying target industries and then supports them through an

elaborate network of collaborative research and development to direct financing.

2. "Imperfect market" as used here refers to the distortions in the market such as economies of scale, learning curves, and large R&D requirements that can restrict the process of perfect market competition. Recognition of such imperfections does not contradict the belief in the power and purity of the market that many economists hold as superior to government intervention, even in markets accepted as imperfect.

3. Robert White, "Technology and the Bush Administration: Moving Beyond the Industrial Policy Debate," Speech to the National Press Club, Washington, D.C., 29 September 1992.

4. U.S. International Trade Commission, *Foreign Industrial Targeting and Its Effects on U.S. Industries, Phase I: Japan*, USITC Publication no. 1437 (Washington, D.C.: Government Printing Office, 1983), 20.

5. Rice is highly protected in Japan and agriculture in Europe, even though comparative advantage has flowed elsewhere. The American political system is particularly vulnerable to special-interest groups lobbying for protection.

6. Peter J. Sprague quoted in Robert Reich, "The Real Economy," *The Atlantic Monthly*, February 1991, 35-52.

7. Martin Libicki, What Makes Industries Strategic (Washington, D.C.: National Defense University, November 1989), 26; "How Boeing Fights Airbus," *The Economist*, 30 January 1988, 51.

8. David B. Audretch, Leo Sleuwaegen, and Hideki Yamawaki, eds., The Convergence of International and Domestic Markets (Amsterdam: Elsevier Science Publishers, 1989), 287. See also Richard Nelson, *High Technology Policies: A Five Nation Comparison* (Washington, D.C.: American Enterprise Institute, 1985).

9. The motivating impetus has always been to negotiate the deal that offers the best reward and to source inputs that offer the cheapest alternative. If that meant licensing technology, selling technology, or shifting assets overseas, then it was the appropriate action to take. Raymond Vernon, "A Strategy for International Trade," *Issues in Science and Technology* (Winter 1988), 86-91.

10. Kenichi Ohmae, *The Borderless World* (London: Collins, 1990), 218.

11. Charles Kindleberger, *American Business Abroad* (New Haven: Yale University Press, 1969); Quoted in Yao-Su Hu, "Global or Stateless Corporations are National Firms with International Operations," California Management Review 2 (Winter 1992), 107-108.

12. Michael Porter, *The Competitive Advantage of Nations* (New York: Free Press, 1990), 19.

13. Bernard Cole, "Smart Memories are Eating into the Jelly Bean Market," *Electronics*, 5 February 1987, 65.

14. Japanese electronic houses see competitive success inextricably linked to the control of their parts manufacturing. "What's Happening to British Chips," *The Economist*, 12 December 1987, 76.

15. Martin Libicki, *What Makes Industries Strategic* (Washington, D.C.: National Defense University, November 1989), 26.

16. Yao Hu, "Global or Stateless Corporations are National Firms With International Operations," *California Management Review* 2 (Winter 1992), 119.

17. Martin Libicki, *What Makes Industries Strategic* (Washington, D.C.: National Defense University, November 1989), 27.

18. *The Economist*, 3 March 1988, 66; cited in Ibid., 27.

19. William Gsand, executive vice president of Hitachi America, states that the best

technology and processes are always stationed at headquarters. There is, therefore, a national stake in maintaining a high technology framework headquartered in the United States. William Gsand, personal interview, 12 August 1992.

20. Public R&D represented 14% of total R&D before the war and to a greater or lesser extent over 50% after the war. See U.S. Bureau of the Census, *Historical Statistics of the United States: Colonial Times to 1957* (Washington, D.C.: Government Printing Office, 1960), 965.

21. The United States spent over half of the government R&D budget on defense-related projects in the 1980's. Japan spent, on average, 2% during the same period. See Kenneth Flamm, "Technology Policy in International Perspective," in U.S. Congress, Joint Economic Committee, *Policies for Industrial Growth in a Competitive World*, 99th Cong., 2d Session (Washington, D.C.: Government Printing Office, 1984), 29-39.

22. The American tax code is replete with provisions that promote some economic sectors over others.

23. U.S. Congress, House of Representatives, *Council on Industrial Competitiveness Act*, 99-579, 99th Cong., 2d Session (Washington, D.C.: Government Printing Office, 1986), 11.

24. Phyllis Levinson, *The Federal Entrepreneur: The Nation's Implicit Industrial Policy* (Washington, D.C.: Urban Institute, 1982), 70, 85.

25. Washington Post, 4 October 1987, H1; Alan W. Wolff, "International Competitiveness of American Industry: The Role of U.S. Trade Policy," in *U.S. Competitiveness in the World Economy*, Bruce Scott and George C. Lodge, eds. (Cambridge: Harvard Business School Press, 1985), 324.

26. Raleigh News and Observer, 17 August 1983, quoted in Otis L. Graham, Jr., *Losing Time: The Industrial Policy Debate* (Cambridge: Harvard University Press, 1992), 191.

27. Otis L. Graham, Jr., *Losing Time: The Industrial Policy Debate* (Cambridge: Harvard University Press, 1992), 206.

28. President Jimmy Carter was less ideologically opposed to the concept and imitated studies on the topic, but did not, with the exception of the synfuels program, enact a government promotion agenda. See Richard Nelson, *High Technology Policies: A Five Nation Comparison* (Washington, D.C.: American Enterprise Institute, 1985), 147.

29. The Trade and Technology Promotion Act of 1989 proposed to set up the Advanced Civilian Technology Agency (ACTA) within the Department of Commerce. The agency would offer grants, and enter into contracts or cooperative agreements to support projects for developing new technology of importance to the United States. See U.S. Congressional Budget Office, *Using R&D Consortia for Commercial Innovation: Sematech, X-Ray Lithography, and High-Resolution Systems* (Washington, D.C.: Government Printing Office, 1990), 18.

The Defense Advanced Research Projects Agency (DARPA) funds and manages research on projects considered important to American defense. The overriding mission of DARPA is to nurture American technological superiority and to make certain that the United States would not be caught off-guard technologically, as it was after the Soviet *Sputnik* launch in 1957. DARPA is the Pentagon's chief organ for picking high technology sunrise industries. In the 1980's, the agency began work on very high speed integrated circuits (VHSICs), advanced lasers, fiber optics, computer software, and composite materials. It has a budget of approximately $1.3 billion. Ibid., 108.

30. European Commission queried why the United States would willingly forfeit a whole industry to the Japanese and suffer the negative trade effects and damage to the electronic food chain that it would probably entail. Ibid.

31. Cynthia A. Beltz, *High-Tech Maneuvers: Industrial Policy Lessons of HDTV* (Washington, D.C.: The American Enterprise Institute, 1991), 16.

32. Charles Ballard and others submit that the marginal efficiency cost of a dollar revenue in the American tax system is between $1.17 and $1.56. See Charles L. Ballard, John Shoven, and John Walley, "General Equilibrium Computations of the Marginal Welfare Costs of Taxes in the United States," *American Economic Review* 75 (1989), 128-138.

33. Anne O. Krueger, "Theory and Practice of Commercial Policy: 1945-1990," *NBER Working Papers* 3569 (Cambridge: National Bureau of Economic Research, December 1990), 106-21; Deborah Spar and Raymond Vernon, *Beyond Globalism: Remaking American Foreign Economic Policy* (New York: The Free Press, 1989).

34. Lester C. Thurow, *The Case for Industrial Policies* (Washington, D.C.: Center for National Policy, 1984), 21.

35. Joseph L. Badaracco and David B. Yoffie, "Industrial Policy: It Can't Happen Here," *Harvard Business Review* (November/December 1983), 96-103.

36. Bruce Scott, "National Strategies: Key to International Competition," in *U.S. Competitiveness in the World Economy*, Bruce Scott, ed. (New York: Norton, 1987), 72-108. See also Bruce Scott, "How Practical is National Economic Planning," *Harvard Business Review* (March/April 1978), 131.

37. Charles L. Schultze, "Industrial Policy: A Solution in Search of a Problem," *California Management Review* 24 (Summer 1983), 6.

38. Linda R. Cohen and Roger G. Noll, *The Technology Pork Barrel* (Washington, D.C.: The Brookings Institution, 1991).

39. Laura D'Andrea Tyson, *Who's Bashing Whom: Trade Conflict in High Technology Industries* (Washington, D.C.: Institute for International Economics, 1992), 194.

40. The Advanced Technology Program (ATP) created under the Bush administration and administered by the Under Secretary of Technology in the Department of Commerce. This program offers grants to companies that seek to develop promising but risky precompetitive technology. Awards are made on the basis of practicality, technical merit, and the ability to exploit the product in a successful commercial fashion. Through 1993 the program had committed $175 million to over 45 projects.

41. George Lodge, *Perestroika for America: Restructuring U.S. Business-Government Relations for Competitiveness in the World Economy* (Boston: Harvard Business School Press, 1990), 14.

42. Herbert Stein, "Don't Fall for Industrial Policy," *Fortune*, 14 November 1983, 78. See also Herbert Stein, *Presidential Economics* (New York: Simon & Schuster, 1984).

43. Robert Z. Lawrence, *Can America Compete?* (Washington, D.C.: Brookings Institution, 1984), 80-83. This point is also argued in Michael N. Cantwell, "Global Competition: U.S. Industry's Hidden Advantages," *Industry Week*, 7 October 1991, 53.

44. Otis L. Graham, Jr., *Losing Time: The Industrial Policy Debate* (Cambridge: Harvard University Press, 1992), 127.

45. Murray Widenbaum and Michael Athey, "What is the Rust Belt's Problem," in *The Industrial Policy Debate*, Chalmers Johnson, ed. (San Francisco: Institute for Contemporary Studies, 1984), 128.

46. Melvyn Krauss, "Europeanizing the U.S. Economy: The Enduring Appeal of the

Corporatist State," in *The Industrial Policy Debate*, Chalmers Johnson, ed. (San Francisco: Institute for Contemporary Studies, 1984), 72-73.

47. The Fulbright investigation of 1951 illuminated extensive waste and fraud. The RFC loaned funds to enterprises from steel companies and hotels to snake farms and roulette wheels. Otis L. Graham, Jr., *Losing Time: The Industrial Policy Debate* (Cambridge: Harvard University Press, 1992), 126-128.

48. The United States and the United Kingdom share a common cultural heritage and a common history of once being dominant economies in the world. Both countries industrialized without dynamic government direction and shared a legacy of holding dear to laissez-faire principles during their relative eclipse from global economic dominance.

49. Richard E. Caves and Lawrence B. Krause, *Britain's Economic Performance* (Washington, D.C.: Brookings Institution, 1980), 94. See also Stephen Blank and Paul Sacks, "If At First You Don't Succeed, Don't Try Again: Industrial Policy in Britain," in *Industrial Vitalization: Toward a National Industrial Policy*, Margaret E. Dewar, ed. (New York: Pergamon, 1982).

50. Bernard Uddis, *The Challenge to European Industrial Policy: Impacts of Redirected Military Spending* (London: Westview Press, 1987), 115.

51. Raymond Vernon, "Enterprise and Government in Western Europe," in *Big Business and the State*, Raymond Vernon, ed. (Cambridge: Harvard University Press, 1981), 13.

52. Brian Hindley, *State Investment Companies in Western Europe: Picking Winners or Backing Losers?* (New York: St. Martin's Press, 1983).

53. Jean-Claude Derian, *America's Struggle for Leadership in Technology*, translated by Severen Schaeffer (Cambridge: MIT Press, 1990). See also John Zysman, "Between the Market and the State: Dilemmas of French Policy for the Electronics Industry," *Research Policy* 7 (1978).

54. Cited in Thomas R. Howell, Brent L. Bartlett, and Warren Davis, *Creating Advantage: Semiconductors and Government Industrial Policy in the 1990's* (San Francisco: The Semiconductor Industry Association, 1992), 43.

55. Theodore Eismeier, "The Case Against Industrial Policy," *Journal of Contemporary Studies* 6 (Spring 1983), 21-22.

56. Phillip H. Trezise, "Industrial Policy is not the Major Reason for Japan's Success," *Brookings Review* 2 (Spring 1983), 13-18.

57. Takatoshi Itoh, *The Japanese Economy* (Cambridge: MIT Press, 1992), 201.

58. MITI has mixed results in picking winners, as illustrated by shipbuilding, commercial aircraft, and the belief in 1960 that Japan should not even attempt to compete with the United States in automobiles. See Yukio Noguchi, "Government Business Relationship in Japan," in *Policy and Trade Issues of the Japanese Economy*, Kozo Yamamura, ed. (Seattle: University of Washington Press, 1982).

59. MITI has been counterproductive in many sectors such as construction, agriculture, petroleum and related products, retailing, and paper products, in which Japanese productivity and quality remain behind the world competition. See Michael Porter, *The Competitive Advantage of Nations* (New York: Free Press, 1990), 665, 708.

60. Jared Taylor, *Shadows of the Rising Sun: A Critical View of the Japanese Miracle* (New York: Morrow, 1983), 21-22.

61. Robert Ozaki, "How the Japanese Industrial Policy Works," in *The Industrial Policy Debate*, Chalmers Johnson, ed. (San Francisco: Institute for Contemporary Studies, 1984).

62. Toshimasa Tsuruta, "The Myth of Japan Inc.," *Technology Review* 86 (July 1983).

63. Richard T. Pascal and Anthony G. Athos, *The Art of Japanese Management: Applications for American Executives* (New York:Warner Books, 1982).

64. Daniel Okimoto, *Between MITI and the Market: Japanese Industrial Policy for High Technology* (Stanford: Stanford University Press, 1989).

65. The LDP's long dominance of Japanese politics fostered consistency and long-term vision, a light military burden, homogenous population, structural features of Japanese industrial groups, low cost of capital, sound macroeconomic policies, and the capacity of MITI to marshal support among the private sector for development strategies in specific sectors. Daniel Okimoto, *Between MITI and the Market: Japanese Industrial Policy for High Technology* (Stanford: Stanford University Press, 1989), 238.

66. Ibid., 231.

67. Hugh Patrick, "Japanese High Technology Industrial Policy in Comparative Context," in *Japan's High Technology Industries: Lessons and Limitations of Industrial Policy*, Hugh Patrick, ed. (Seattle: University of Washington Press, 1986).

68. Ken-ichi Imai, "Japan's Industrial Policy for High Technology Industry," in *Japan's High Technology Industries: Lessons and Limitations of Industrial Policy*, Hugh Patrick, ed. (Seattle: University of Washington Press, 1986), 141.

69. Richard Nelson and George Eads, "Japanese High Technology Policy: What Lessons for the U.S.?" in *Japan's High Technology Industries: Lessons and Limitations of Industrial Policy*, Hugh Patrick, ed. (Seattle: University of Washington Press, 1986), 263.

70. Ibid.

71. Hugh Patrick, "Japanese High Technology Industrial Policy in Comparative Context," in *Japan's High Technology Industries: Lessons and Limitations of Industrial Policy*, Hugh Patrick, ed. (Seattle: University of Washington Press, 1986), 29.

72. Porter submits that research and development "cannot be left solely to firms due to spillovers that not only benefit the firm but often raise the rate of advancement in the entire national industry." Michael Porter, *The Competitive Advantage of Nations* (New York: The Free Press, 1990), 631, 624.

73. An exception may be in aerospace. The European Airbus knocked McDonnell Douglas out as the second largest manufacturer and has made substantial inroads into the once enormous world market share of Boeing. Jean-Claude Derian argues that this has been a successful industrial policy because it was oriented toward global competition and not sheltered from competition (reason for failed national champion policies in electronics). Jean-Claude Derian, *America's Struggle for Leadership in Technology*, translated by Severen Schaeffer (Cambridge: MIT Press, 1990).

74. Christopher Freeman, *Technology Policy and Economic Performance* (London: Francis Pinter, 1987).

4

Strategic Trade Policy

Global economic integration has pried open national markets to foreign competition, and as a result, no industrial policy may be conceived without reference to related trade issues. The success of any enterprise, particularly those required to make large front-end investments in R&D, depends upon the widest access to markets to cover production costs. Distortions in free trade that include export subsidies, tariffs, quotas, and other instruments of state-led development efforts are important features of industrial policy. They are also less tolerable to an American government that, on the one hand, recognizes its declining hegemonic power and, on the other, is unsure of the practical utility of these policy tools for economic strength. The official trade policy of the United States has been based on principles of free trade. However, there is growing uncertainty as to how to reconcile violations of liberal trade with national interest in the global economy. It is an issue that has gained in importance as traditional economic theories have not adequately explained emerging patterns of international competition.

Orthodox trade theory (classical theory) is based on models of perfect competition and examines trade patterns in light of comparative advantage reflected by fixed factor endowments of land, labor, and capital. It holds that competitive forces are most efficient in allocating resources and prices to an equilibrium, and consequently there is little allowance made for the possibility that government intervention could efficiently reshape private economic activity and alter the concept of comparative advantage. Technology was an exogenous factor, and as such the classical theory provided little schematic room to explain the proliferation and development of brain-power industries not dependent upon traditional factor endowments. Since high technology industries can be located wherever the best technology is developed and applied, national policies may influence the formation of technological capital and conditions of comparative advantage. This new orientation in international economics recognizes that technology is a central factor in international trade, and implicitly considers those deviations from perfect market competition that strategic trade policy advocates argue can be manipulated to shift the comparative advantage of strategic industries.

NEOCLASSICAL TRADE THEORY AND THE NEW ECONOMICS

The principal architect of the classical theory of international trade was David Ricardo. His premise of comparative advantage as the fundamental engine of international trade held that countries trade according to their relative strengths. Factor endowments (land, labor, capital, natural resources) determined what they could produce most efficiently and export accordingly.[1] This "factor proportions theory" stipulated that trade patterns and industrial organization are shaped by the relative abundance of the factors of production. Countries capital rich will export capital-intensive products while countries capital poor or labor rich will concentrate on those goods that utilize their comparative advantage in labor. Trade models were assumed to be characterized by constant returns and perfect competition, and international trade occurred because of the differing tastes and factor endowments among countries.

The classical theory regarded free trade as the best policy. Comparative advantage dictates that trade generates gains by permitting specialization among countries and this mutual benefit is occasioned by the principles of the competitive market structure. In perfectly competitive markets, the appropriate amount of trade will be undertaken, and the most efficient way to allocate scarce resources such as capital, skilled labor, and raw materials is to allow the market to work unfettered. There are no specific decisions on what should be produced and how; rather, firms set certain priorities that reflect the conditions set by the free market. Since international markets are not seen as different from domestic ones, the argument is that we should allow free trade to allocate the productive sources even if not all countries in the international system practice free trade.

This theoretical characterization of international trade posed problems when applied to real-world policy situations. Even though free trade may be good for the world as a whole, individual countries may be able to gain unilateral advantages from interventionist policies. In addition, any policy change, including trade liberalization, creates winners and losers. While free trade disperses its gains over the whole consumer population, losers tend to be concentrated in specific sectors and are thus more easily identified. Even though a policy of free trade may generate far more gains than losses, the potential losers often pose a more coherent and stronger political force to block the policy.

Since 1945 a growing percentage of trade has been occasioned by exchanges that cannot readily be attributable to the concept of comparative advantage. Most world trade takes place between advanced nations with broadly similar factor endowments. International trade competition is increasingly determined by a series of temporary advantages resulting from economies of scale or shifting leads in technological races, and technology is increasingly viewed as the engine driving specialization and comparative advantage. In many industries competitive advantage appears less determined by factors of endowment and economies of large-scale production than by knowledge generated by firms through R&D and experience.[2] In this regard, access to abundant factors is less important than the technology and skills required to produce goods or services efficiently.

Orthodox economic theories recognize that monopoly or monopsony conditions do exist, but for the most part, they assume that world markets are basically "perfectly competitive" where many producers are incapable of individually affecting market prices or the actions of their competitors. However, it is clear that there are industries in which there are few firms, where individual firms have the power to affect equilibrium market prices and face few identifiable rivals whose actions they can influence. Firms in this framework are called *oligopolies* and operate in what is termed "imperfectly competitive markets."

A reconsideration of traditional theories is appropriate because trade based on  simple concepts of comparative advantage, if it ever really existed, has been displaced by trade based on a more complex set of factors. Simple Ricardian and factor proportions theories assumed constant returns to scale and perfect competition, while ignoring or dismissing phenomena relating to externalities, R&D, economies of scale, learning curves, and market evolution. Furthermore, comparative advantage principles were not dynamic and did not allow for the possibility that such comparative advantage could change over time and be manipulated by an active trade policy. Bruce Scott addresses this point directly:

> The challenge to traditional theory comes first and foremost from evidence that a number of countries have indeed succeeded in dramatically altering their comparative advantage. . . . With the advantage of hindsight, it is obvious that the short-term and long-term growth and productivity prospects for cloth and wine were quite different in the original Ricardian example. For Portugal, the short-term advantage comes from specialization in wine; the long-term advantage comes from making a success in cloth, the "high tech" high-growth, rapidly changing industry of the period. . . . In short, the Portuguese should have specialized in textiles, not cloth, regardless of whether their costs were lower or higher than those prevailing in Britain at the time.[3]

Because factor endowments were fixed, technology was exogenous,[4] and constant returns to scale were assumed, the classical theories could not address the issue of creating a workable comparative advantage. Traditional theories left little scope within which to measure trade policy as a means to improve a nation's welfare. The single exception to this rule was the recognition of the "optimum tariff." This conceded the possibility that a large country could, by restricting free trade, exploit monopoly or monopsony power in international markets, provided such power was not exploited by competing domestic trade structures.[5] Even this, however, was shown to be self-defeating if the trading partner assumed a retaliatory position.

The theoretical and empirical shortcomings of traditional economic theory prompted a reassessment of the predictive value of existing international trade theory. What emerged was a strategic trade policy school that justified the mercantilist arguments for restricting imports and promoting exports on the premise of profit shifting.[6] The new trade theory emphasizes increasing returns and imperfect competition, and it questions to what extent current international trade can be explained by the traditional Ricardian concepts of comparative advantage. The notion that increasing returns are a cause of specialization and international trade is not new, but it has not been until recently that this has been successfully plugged into a model of market structure.

Classical trade theory could not account for externalities in its models, as the know-how investment of firms that is the source of the externality is not consistent with the paradigm of perfect competition. The more goods a company can produce the cheaper each successive product becomes because most of the cost is the front-end investment in R&D. Because the immediate investment in know-how pays off with lower unit costs of production, increasing economies of scale, and thus the dynamic economies of scale, must lead to a breakdown of perfect competition. The firms that are first to go to market are first to move down the learning curve and scale economies and will, therefore, be in a position to underprice tardy competition. The presence of externalities, while recognized, was removed from operational utility by orthodox trade theory.

The notion that protection can benefit an industry that generates external economies is part of the traditional theory of trade policy, although this is a second-best policy as conventionalists regard the correction of domestic market failure as preferable. The new trade theory suggests, however, a greater role for government intervention to promote these external benefits. The new models argue for the restriction of imports and promotion of exports to effect a profit shift to the domestic producer. The concern that rival governments could capture permanent advantage in industry after industry by giving each a small initial impetus down the learning curve emerged as a primary consequence of this new orientation. New models have given the notion of increasing returns as an impetus for trade a new level of clarity and precision. International trade theory has become fused with the concept of industrial organization.[7]

If we assume that economies of scale and imperfect competition are the norm rather than the exception in many high technology industries (such as semiconductors or aerospace), the abstractness of applying this to a model is reduced. Since external economies can be associated with investment in knowledge (R & D), it is implied that those firms engaged in extensive R & D as a proportion of the firm's costs are most likely to have these external economies. The difference between government's targeting of industries with high external economies and the strategic trade policy described above is that the promotion of domestic externalities need not adversely affect a foreign competitor.

New analysis suggests that active intervention can benefit a country relative to free trade at the expense of a competing nation. This is possible by using government policy to secure larger rents,[8] and by using government intervention to garner increasing external economies. If there are advantages to large-scale production or a steep learning curve, new entry into an industry could appear unprofitable even if existing firms are enjoying high profits. It is then possible to use government subsidy or protection to increase the rents accrued to the home firm at the expense of a foreign competitor.

While both free trade and strategic trade policy advocates regard the maximization of national welfare as the primary objective, the strategic trade policy seeks to gain national advantage at the expense of a foreign competitor. Orthodox trade theory conceded a policy conflict only in the optimum-tariff case. The difference is that the new analytic current strives to determine if changes in the allocation of resources matter, an objective premised on there being strategic sectors in the economy where capital and labor receive a higher return than they could in an alternative economic activity. Orthodox economic theory indicates that there are no such strategic sectors, that competition will eliminate any distortions that could create such a sector. By extension, market prices that guide the allocation of resources reflect true value.

STRATEGIC TRADE POLICY

Increasing returns can be an independent cause of international specialization and trade. New models recognized two new trends in the postwar trading system. The first was the emergence of previously unindustrialized countries of the Pacific Rim, reaching a level of industrial activity comparable to the largest postwar powers. (the NICs, or newly industrialized countries). The second is the conglomeration and multinationalization of corporate enterprises. The two combine to create a trading environment with small numbers of large, interdependent firms and governments and away from a trading system composed of large numbers of small, geographically separated, quasi-independent firms and governments.

Strategic trade proponents argue that theories of externalities based on perfect competition neglect the source of the externality, since large fixed costs are involved in R&D and large fixed costs keep the number of competitors in an industry to a minimum. There are three key market imperfections that potentially can be exploited by government intervention: large economies of scale, steep learning curves, and sizable R&D requirements. The first two conditions are quite important to the firm as they create the possibility of receiving increased rents. If the production of goods and or services in an industry has large economies of scale or substantial learning effects, then access to foreign markets and the behavior of foreign firms and governments directly affect the profitability of the domestic industry. Under these conditions, corporate profitability at home becomes interdependent with the actions of other countries.

Since one or more of these imperfections can cause barriers to entry over time, foreign protectionism or subsidies can potentially give foreign competitors cost advantages that later entrants could not match. In an industry in which the U.S. market is open and a large foreign market is closed, foreign competitors could achieve more efficient scale as a result of increased domestic and overseas sales, while domestic producers would be squeezed into a portion of the domestic market.[9] Once firms in an industry fall behind, it would be difficult to recover profitability.

The new models accommodate aspects of the modern international trading system that had previously been omitted by traditional models. In large measure, the new international economics was developed by applying the tools of the "new" industrial organization in an international context.[10] The new models are broadly consigned to two classes of models. The first is models that explain intra-industry trade. This implies that the production and traded goods are not associated with rents or externalities that are open for capture by other governments. It does not matter, therefore, if a country has an industry in which it specializes in the production of one good or another. Import protection can occasion export promotion.[11] The second category is that of strategic trade policy models. These models assume that there are rents to be captured and maintain that increasing the domestic production of some tradable good means increasing a country's share in potential monopoly rents or external benefits associated with this activity. Both categories of models insist that government intervention can benefit the national welfare and endeavor to determine the optimum form such intervention might take. The models analyze policy behavior of a small number of countries acting as agents in support of domestic producers in an international oligopolistic industry.

Brander-Spencer Profit-Shifting Model

The models pioneered by James Brander and Barbara Spencer formalize the premise that government can increase national welfare by promoting domestic development of industries that create substantial factor rents or externalities.[12] This model suggests that an active government policy can increase national welfare at the expense of another country. This could be achieved via subsidy or import protection. For example, if there is a large domestic market for a good, protectionist measures can raise the profits of the domestic firm, and lower the profits of the competing foreign firm; and like an export subsidy, this protection would have the effect of deterring foreign entry and allow the domestic firm to capture the excess returns. The objective is to shift monopoly profits from a foreign enterprise to a domestic concern. Such monopoly profits are assumed to occur for an international industry because entry is restricted.

Brander and Spencer assume the existence of an industry consisting of two firms making a homogeneous product. The two firms are located in two different countries and serve a common export market in a third country where all sales and profits are initially made. They can then equate an increase in a national producer's profit (minus subsidies) with an increase in national welfare. The problem of profit shifting, then, is best resolved by analyzing how intervention by the government of one of the producing countries can increase the profit that its domestic producer extracts from the duopolist's common export market.

In the Brander-Spencer model each firm uses output as its action parameter and both are assumed to behave like Cournot duopolists.[13] In the absence of government intervention, this equilibrium corresponds to the Cournot solution. That is, equilibrium will occur when each firm chooses the level of output that is optimal given the output of its rival and that neither will have enough incentive to alter its behavior. However, should the American firm choose to expand sales in the foreign country, the result would be a fall in price for the good, the American firm would gain increased market share, and the firm would offset the foreign competitor and induce him to restrict output. The foreign firm could see this as a bluff since it assumes that the American firm is producing at an optimal level.[14]

Herein lies a possible role for government. An export subsidy is issued to the American firm, the bluff is made more credible, and the foreign firm is obliged to contract output. In this scenario government action would alter the competitive climate and push the market to a new equilibrium in favor of the American producer. The increased production further expedites the American firm's move down its learning curve, to gain experience and become more efficient.

A Stackleberg solution dictates that a sophisticated duopolist who anticipates the rival Cournot adjustment to his own changes in output can do better than a Cournot duopolist.[15] If the domestic firm produces more than the Cournot output equilibrium, it can push its competitor down its reaction function, increasing its own profits at the expense of its rival. This profit enhancement can occur if the domestic firm assumes the role of a Stackleberg leader and if the rival is content to take a Cournot duopolist's dependent supply position. This leadership position necessitates that there be a credible commitment shown. In this model they consign the government to the role of inducing the promoted firm to produce the volume that corresponds to the Stackleberg leader position. The foreign government is assumed to remain passive. While firms act like Cournot duopolists all the time, the government manipulates the rules of the competition and the subsidized firm enjoys increased sales and decreasing per-unit costs. The Stackleberg leader garners increased market share and moves down its learning curve, reaping production efficiencies. Brander and Spencer acknowledge that an important assumption in this model is that "the government understands the structure of the industry and is able to set a credible subsidy on exports in advance of the quantity decision by firms."[16] The logic here is that only government can make a credible commitment to maintaining the leadership position and for this reason Brander and Spencer justify government intervention in their model.[17]

Even without intervention, each duopolist would have the incentive to occupy the position of a Stackleberg leader if its rival would be content to act as a follower. Moving from Cournot equilibrium to a leadership position is, however, not considered a credible choice for either firm acting on its own because the other duopolist cannot be expected to reduce its output. In the absence of government intervention, the two firms are assumed to be on equal footing. Each understands that an expansion in output will not be profitable if the other does not retreat as an export subsidy alters the situation. Expansion of the subsidized firm's output would then be regarded as a credible choice by its rival because an expansion would be privately possible even if the rival would not reduce its output. The American firm would find it profitable to expand sales aggressively into the foreign market even if the rival firm maintained its present output because of the subsidy. In the eyes of the foreign rival, the government intervention has to be seen as credibly committed to the intervention. The foreign firm would have to believe that the subsidy would continue even if it did not contract.[18]

To close the policy credibility issue two conditions present themselves.[19] First, the subsidy, by lowering the firm's marginal cost, has the effect of shifting the subsidized firm's reaction curve outward. With both duopolists moving along their reaction curves, a new Cournot equilibrium is attained at an increased level of output for the subsidized firm and at a reduced level of output for the foreign rival. Second, this contraction by the rival increases the price that the domestic firm can obtain for any given level of output and causes profits to rise through that channel. The subsidy has two effects. The first effect is a cost saving to the promoted firm and the second is to encourage the rival firm to contract its production activities. This second and strategic effect raises the profits of the promoted firm further by selling additional output, and the net effect is that the gains accrued to the subsidized firm exceed the cost of the initial subsidy.

This is a predatory policy subsidy as the gains accruing to the domestic enterprise occur at the expense of the foreign competitor. However, from the point of view of the global consumer, this process is welfare enhancing as it provides the same product at a lower cost, as the subsidy acts to reduce the monopoly distortion in this imperfectly competitive industry. In a welfare analysis, therefore, Brander and Spencer would argue that the benefit to the firm exceeds the cost to the taxpayers.

Analysis of trade policy based on competitive models offers little support for a trade policy activism of any sort and none for export subsidies, which are generally seen as distorting allocation of resources and tend to worsen the subsidizer's terms of trade. In contrast, the Brander-Spencer model presents a model of imperfect competition that represents a reversal of the conventional wisdom of trade theory. Export subsidies may not be the best policy, but at the very least they are shown to be at least a reasonable policy instrument.

Brander and Spencer admit that the model is not adaptable to changes in assumptions and submit that the basic point is more generally applicable than the specific model.[20] Government action can alter the strategic balance and assist domestic firms in international competition. It is not a clear-cut case for the employment of across-the-board subsidies for export, but rather it is a policy in which one would have to target particular industries—pick winners—which is not easy. The model illustrates how a strategic subsidy might work. In profitable markets domestic firms are made better off if foreign firms can be induced to contract, or to expand more slowly than they otherwise would have.

Krugman's Model: Strategic Infant-Industry Policy

Trade policy has traditionally concerned itself with the protection of domestic import-competing industries rather than export promotion. The main tools for this policy have been the tariff and the import quota. Paul Krugman presents a model that adds a strategic dimension to the old infant industry argument for government intervention.[21] The infant-industry argument is that temporary protection of an industry that cannot currently compete with foreign rivals might be justified if that industry, through protection, can survive and become more efficient and eventually compete with those same foreign firms.

Krugman redirects the focus of infant-industry analysis largely by making unorthodox assumptions (i.e., that a country has an opportunity to achieve a different kind of welfare gain through the strategic deployment of import restrictions). The model's logic depends upon two basic components: the presence of international oligopolies and economies of scale.[22] With respect to oligopoly, Krugman's model assumes the existence of two firms, one domestic and the other foreign. Each firm produces a single product that it sells in several segmented markets, for example, domestic and foreign markets. The products of the two firms are close substitutes but they need not be perfect substitutes. Like Brander and Spencer, Krugman assumes that both firms at all times act as Cournot duopolists. The result is a "multi-market Cournot model."[23] Regarding his second basic component, Krugman distinguishes three different forms of scale economies: marginal cost, restriction of a market, and learning by doing.[24]

The idea behind learning by doing is that as a firm produces more it learns how to undertake further production more efficiently. By giving the domestic producer a privileged position in the home market, a country gives it an advantage in scale of production over a foreign rival, an advantage that translates into lower marginal costs and higher market share even in unprotected markets. The strategic effect of learning by doing is that when one government excludes the foreign producer from a market previously open to it, the intervention causes opposite effects on the two rivals. The domestic producer will sell more in the market from which the foreign rival has been excluded, and thus the domestic firm's marginal cost will fall. The foreign firm's marginal cost will rise as it produces less when excluded from the protected market. Both firms are then further induced by opposite changes in their marginal cost to adjust sales also in unprotected markets.

The domestic firm will expand its output while the foreign firm will be required to retreat further. Again, these adjustments have opposite effects on each firm's marginal cost and the process continues until a new multimarket equilibrium is reached The essential feature of the model is the circular causation from output to marginal cost to output, and it is this circularity that makes import protection an export promotion device. Krugman does not provide a national welfare analysis for his model of strategic trade policy, but it is clearly implied in his argument that additional exports are profitable for the domestic producer. Export prices must exceed marginal cost and the expansion of exports must result in a higher absolute profit from foreign sales. It is, therefore, conceivable that a country's welfare gain from additional profits on exports and lower costs of output sold domestically exceeds the loss of consumers' surplus caused by protection of the home market.

In practice, the Krugman strategy is simply the Brander-Spencer profit-shifting model achieved by other means. Indeed, Krugman employs the same formal apparatus as Brander and Spencer save for his assumptions concerning scale economies and market segmentation. Home market protection moves out the domestic duopolist's reaction curve into the unprotected market, but it also shifts in the foreign duopolist's reaction curve.[25] The result is a new Cournot equilibrium in which the domestic firm gains a greater market share in the export market. This happens for two reasons: its reduced marginal cost has made the expansion attractive, and the foreign rival is willing to retreat. By symmetry, the foreign firm has two reasons to retreat: its rival's credible expansion and its own higher marginal cost. As in the Brander-Spencer model, the strategic rationale of government intervention is that intervention induces a profitable expansion of export sales when it would not be credible for any duopolist to attempt such an expansion on its own.

Rapidly moving down the learning curve in the Krugman model motivates the government intervention because of its strategic effect. The domestic firm extracts a higher profit from an export market because a foreign rival makes room for the government-supported firm. The general point is that trade patterns and potential gains from trade are more sensitive to policy intervention if one removes the orthodox assumptions of constant returns to scale and perfect competition.[26]

IS THERE A POLICY PRESCRIPTION?

That international trade has increasingly become a function of dynamic economies of scale and imperfect market conditions has been readily accepted by most in the economics profession, and in large measure, the debate today reflects disagreement over the appropriateness of government intervention and the problems that inhere to such an active trade policy.[27] At the same time, trade economists cannot simply return to the previous policy posture even if the current models of strategic trade policy are found wanting as blueprints for policy action. The new theories of international trade economics are regarded by some as either too incomplete or at least insufficiently developed to serve as a basis for policy advice.[28] Strategic trade policy demonstrates the possibility that a government, under certain conditions, can improve the national welfare by shifting profits from away from foreign producers. However, translating that into a policy action is highly problematic given real-world situations.

This new orientation has energized international trade theory by infusing ideas from models that formerly were used primarily in the field of industrial organization. This revitalization has manifested itself in a rich menu of theoretical and empirical research. This research more closely addresses the problems that concern individual firms, industries, interest groups and policy makers in the real world. Moreover, as economists undertook to focus on questions that had previously been excluded by the assumption of perfect competition, there appeared to develop a new consensus concerning the strategic potential of activist trade policy or activist industrial policy.

On a theoretical level, the conclusions of strategic trade policy are rather fragile.[29] The Brander-Spencer model is only as strong as the assumptions on which it is based. The introduction of several realistic assumptions that are excluded from the model can render the conclusions more ambiguous and lead to the view that government intervention may be nugatory, or worse, actually harmful. The criteria for selection and information required to implement a successful intervention are daunting, and it becomes questionable whether government has sufficient information to implement successfully an active trade policy.

Under the Brander-Spencer assumptions an export subsidy should be used to shift oligopolistic profits to domestic firms. This line of argument for export subsidies depends crucially on the Cournot quantity setting: each firm fixes the quantities it will produce and offer for sale, and the prices adjust to clear the markets. However, given a Bertrand competition setting in which both firms fix their prices and then produce enough to meet the demand that comes their way, the strategic trade policy implied is just the opposite. If the foreign firm does not respond to the American subsidy, then export promotion would indeed serve the national interest. However, let us assume Bertrand competition instead of the Cournot equilibrium setting. If each firm takes the other's price as given, an export subsidy, instead of reducing competition through its deterrent effect, will in effect intensify the price war and thus raise an exporter's profits by less than the amount of the subsidy. A government can lend credibility to a higher price for its firm if it levies a tax on its sales in markets where it competes with its foreign rival. Then the foreign firm would recognize the higher price as credible, and would respond by raising its own price to mutually profitable level. The government could lend credibility and ensure a higher price for American exports by adopting an export tax as its optimal trade policy. Choosing the appropriate policy tool becomes highly problematic given the ambiguity of firm behavior. Uncertainty over how the firms in the industry will actually respond indicates that the adoption of one policy could actually be harmful should the competing firms act in a way incongruent with presumptions of their behavior.

Any type of government intervention thus potentially becomes welfare reducing if one assumes that the home firm's conjectures are consistent in the sense that its beliefs about the foreign firm's response to its own actions are borne out exactly by the foreign firm's actual response. A domestic duopolist needs no prompting to assume the position of a Stackleberg leader if it correctly anticipates that its foreign rival will act as a follower in the absence of intervention. No profit shifting is possible and free trade is optimal. Government intervention is not indicated either if the foreign rival proves unwilling to make room for the domestic firm's expansion in spite of state support. Thus, even if the structure of the most simplified Brander-Spencer model is assumed, the conjectural element of the oligopolistic relationship makes it practically impossible to extract a reliable policy prescription.

The argument for strategic export promotion rests on the premise that there are above-normal profits and excess returns to be cultivated in a given industry. However, such returns may be temporary owing to the entry of new rent-seeking firms that will compete away excess returns. The entry of new players in the industry is most likely to originate in the country subsidizing the industry, and this can adversely affect national welfare in two ways.[30] First, the promotion can encourage excess entry into the industry that would lead to higher average costs of production in the long-term than would have been the case without the subsidy. Second, many domestic firms entering into the industry will compete with one another thus driving down the export price of the good, thereby compromising any long-run gain envisioned by the export subsidy. Moreover, even if such intervention succeeded in dissuading a foreign competitor, it is argued that the resulting gains will be passed along to foreign consumers rather than securing excess returns for the domestic producers.[31]

Once the welfare of domestic consumers is brought into the picture the ambiguity of the Brander-Spencer model is increased. The Brander-Spencer prescription for profit shifting was derived under the simplifying assumption that the good in question was not consumed in the acting country. Consumption of the exported good in the domestic market will be suboptimal as implied by imperfectly competitive market structures. Export subsidies encourage firms to export goods, thus driving up domestic prices of the same good and exaggerating existing market distortions associated with imperfect competition.

Critics of strategic trade policy models have pointed out that a policy designed to shift profits from foreign to domestic producers can increase national welfare only to the extent that the affected domestic producers are domestically owned.[32] This condition is becoming more questionable every day as we witness the spread of multinational firms, international joint ventures and partnerships, particularly in industries that are typically considered targets for strategic intervention. However, the nationality of the shareholders may be of secondary importance. An important consideration is that the domestic enterprise retain effective technical control, in which case the home nation will still reap many of the benefits to its economy even if the firm is owned by foreign investors or by a foreign firm.

Implementation of strategic trade policies requires the government to identify strategic sectors. Developing rigorous criteria that minimize definitional confusion and serve as a basis for selection is very difficult. Identifying industries in which above-normal profits are earned is a risky endeavor. It is not a simple case of examining profit and wage rates over a specified period. Locating an industry in which the return on labor and capital is exceptionally high can be clouded by many factors. These include the need to separate rent from qualitative differences as higher skilled workers earn a higher return than unskilled labor; the need to determine if technological innovations that lead to a sudden increase of profits and wages in one industry are cyclical or permanent; and the need to treat an industry in light of its successes and failures to get an accurate picture of intrinsic worth. This includes measuring the profits or losses of those firms that are in the industry but do not actually reach the marketing stage of development.

Policy makers would have to apply some subtle reasoning that would require practically unavailable information to select those industries that will create above-average rents in relation to their requirements of the fixed factor margin.[33] Furthermore, external economies are hard to measure. Any spillover effect is extremely hard to quantify in economic terms. The Brander-Spencer model for export promotion applies only to a highly structured scenario of true natural oligopolies where the opportunities for entry at all stages of competition are limited, and in the real world the number of such industries is so small that it removes these models from operational utility.

The authors of strategic trade policy models have thus had a difficult time identifying real-world situations in which to apply their policy prescriptions. Seeming excess profits of firms that are protected by entry barriers, such as patent rights, goodwill, or economies of scale, may in fact represent a risk premium required to reward risky large-scale investments that firms had to make to retain those very patents, goodwill, or economies of scale. If entry barriers are insignificant during the early stages of development, returns adjusted for risk will tend to be normal over the lifetime of the investment. If the government has a policy of targeting successful firms in Brander-Spencer-Krugman fashion, the prospect of subsidization will attract more investment at the early stages than would be attracted without a policy. As one critic pointed out, "this creates a distortion of resource allocation akin to that ascribed to export subsidies in a fully competitive world, because the appropriate long-run view of the industry would indicate the excess profits to be captured by one country or the other by means of strategic policy are nonexistent."[34]

If one were to succeed in identifying a strategic sector, then one must consider the possibility of adverse secondary affects in the "picking of winners."[35] Notwithstanding the data dilemma, the process of picking winners becomes more problematic when you consider if several potential target industries compete for the same factor of production for which supply is not perfectly elastic over the relevant period. The result under an export promotion policy in a targeted sector would be to encourage the promoted industry to expand at the expense of the other sector. The scarce factor required by other sectors is bid up, and these sectors will have to cut back production and yield market share to a ready foreign competitor. Interventionist policies promoting one sector over another must draw resources away from another sector in the domestic economy. The profits of the American firms in non-targeted sectors may fall, potentially offsetting the putative economic gain in the targeted sector.

Government policy could end up encouraging the wrong things as intervention could possibly do more harm than good. The original profit-shifting models ignored this aspect because in the established tradition of partial equilibrium analysis, they focused on one oligopolistic industry in an otherwise perfectly competitive economy. The excess returns garnered by the promoted sector would thus be offset at least partly, by compensating losses elsewhere in the economy. It becomes very difficult to determine the full impact of the policy on the targeted sector and the other sectors that may suffer owing to the intervention. To judge the efficacy of the policy in assisting one sector over an indeterminate side effect on another thus becomes intrinsically difficult.

If there is not one but several American competitors in the market there will be a monetary externality as the American firms compete with each other, overinvest in capacity, and realize a price lower than if they acted collectively in the American interest. Although an export tax would encourage American firms to reduce their output and act less aggressively toward one another, the result would be a loss of some sales to foreign competitors. This same effect can be applied to the issue of external economies. To promote sectors generating external economies draws resources out of other sectors. While there may not be an even ratio between small numbers of competitors and excess returns, or between high R&D expenditures and technological spillovers, there is certainly a correlation that advises caution for any government interventionism.

There is no widely accepted policy action, but there are three possible policy outcomes. First, one country could protect its home market and gain advantage over its international competitors while other nations do nothing. The second case could be that all nations assist domestic firms via protection, with the result being that no firm can do well in the export market. All countries would thus end up with low levels of competition and high costs, as no one firm is able to realize the advantage of size. The third possible outcome is that all countries refrain from protection and resist the temptation to secure unilateral advantage. Unilateral predatory policy is potentially attractive provided other countries remain passive. However, should many countries adopt active policies, then mutual nonintervention would yield the highest combined total return for each country.

Given the theoretical and empirical shortcomings of strategic trade policy, the question of whether political economy concerns will outweigh the benefits of an uncertain policy action becomes important. Strategic trade policy suffers the same political challenge to implementation as discussed for industrial policy in the preceding chapter. In an international context, what would the response be by other countries should the United States adopt an activist trade policy? It could enhance the ability of the United States to change foreign policies detrimental to the United States. It could also undermine cooperation in the international arena, invite retaliatory measures, and lead us into "beggar thy neighbor" trade policies. Politically, strategic policy in support of any individual industry cannot, in practice, be separated from policies concerning other industries. For example, the United States has linked European support for Airbus to other contentious issues, such as grain exports.[36]

Should the trade policies succeed in their objectives, they will have a "beggar thy neighbor" character that may invite retaliatory measures that could ignite a trade war. The economic gains of strategic trade disappear if foreign trading partners respond in kind. If, for example, the English and Japanese react to higher U.S. tariffs on radios by enacting policies to help their own radio industries, the result may be a trade war in the radio market, with ever higher tariffs and subsidies that end up making all three economies worse off. It is, therefore, clear that while models of strategic trade policy can predict the gains and losses that a proposed policy would cause if foreigners responded in one way or another, these same models cannot foresee all the possible ways in which foreign governments or firms might respond.

REPRISE

The composition and direction of trade among the developed economies is increasingly determined by dynamic economies of scale and imperfect market conditions. It is an orientation that lends itself easily to the concept of industrial organization. One formidable task is to determine a method to sort out situations in which government intervention can be justified as in the national interest and situations in which it cannot. Formal economic models still cannot account for the technological linkages and dynamic effects that are stressed by industrial policy advocates. Indeed no theorist could hope to refute the possibility that a country might gain by engaging in a policy of targeting particular sectors or products. However, strategic trade provides an incomplete policy prescription. Even Krugman believes the assertion that high technology sectors generate excess returns is not supported by direct evidence, and he therefore doubts that the gains from strategic trade "of getting there first can mean a big difference in living standards."[37] However, the judgment of any policy must also consider its motivating force. Billions of dollars pumped into the European Airbus Consortium may or may not ever pay dividends in a monetary sense.[38] However, the operating factor was the decision by participating European countries that the maintenance of a viable aerospace industry was strategically desirable.

One critical implication of strategic trade policy is the role it plays in response to another country's trade policies. Traditional trade theory makes it clear that free trade is the best policy and that retaliation almost never makes sense. Strategic trade theory suggests that another country may shift profits to domestic producers. Hewing to an absolute trade line is unwise as foreign countries may believe that they can manipulate trade with impunity.[39] It may or may not pay to initiate an interventionist trade policy, but a country should be prepared to strike against foreign intervention designed to manipulate advantage.

Strategic trade models may appear incomplete as a policy prescription but an increasing number of countries are coupling infant-industry protection and subsidy to secure international advantage in semiconductors. Japan (and other nations on the Pacific Rim) and Western Europe, for example, have not relied exclusively on the invisible hand to guide the commercial development of national semiconductor and other high technology industries. The objective of industrial policy is not necessarily consumer welfare.[40] The Japanese government did not weigh the costs in foregone consumer welfare or the opportunity costs imposed on other sectors because they were considered less important than the creation of a competitive semiconductor industry. For example, without trade protection there may not have been any Japanese producers of 16K chips between 1978 and 1984 because the chips could have been more cheaply purchased from American producers, and the existence of implicit trade barriers resulted in a less than optimal allocation of economic resources.[41] At the same time, indications that it was welfare reducing are not conclusive because a static approach—one that measures extra profits on 16k chips against the added cost of purchasing cheaper versions from American firms—ignores the dynamic effects of a Japanese position in the industry and externalities captured by future generations of semiconductor products and systems incorporating semiconductor technology. Where strategic policy welfare gains are disputed, there are cases in which the contention that policy resulting in a net loss in economic welfare remains questionable. One observer commented that:

> The danger that a significant degree of monopoly power can be acquired and exploited in high technology sectors as the result of strategic policies, endangering the ability of otherwise competitive industries dependent on the targeted sector to compete internationally, is simply too great, and the possible costs for American economic performance too onerous to be ignored.[42]

The supporting trade theory has given new momentum to industrial policy proponents. It has long been accepted that industrial targeting can alter the mix of industries within a given country, but demonstrating that targeting can make an absolute addition to the general welfare of a nation has been more difficult, and the new international trade theories have made a contribution in this regard by providing a model that details how government intervention might increase national welfare. Manipulating comparative advantage has been shown, at least in theory, to be potentially welfare enhancing, and export subsidies may not be so much a gift as a Trojan horse. Orthodox economic theory regards export subsidies as not cost-effective. Subsidies by country A lead to cheaper imports for country B, leading to a contraction of the domestic industry in country B; but the benefits to consumers in cheaper imports outweigh any loss by producers. Strategic trade has shown how subsidies, when viewed strategically, may be less benign than orthodox economics would suggest. However, translating theory into practical policy is so replete with pitfalls that its utility as an industrial policy instrument is doubtful. Yet even if the practical value of the new trade theory is incomplete, it has contributed to the debate on the best role for government in dealing with issues of competitiveness.

NOTES

1. David Ricardo, *On the Principles of Political Economy and Taxation* (London: John Murray, 1817).

2. Richard Baldwin and Paul Krugman, "Industrial Policy and International Competition," in *Trade and Policy Issues and Empirical Analysis*, Richard Baldwin, ed. (Chicago: University of Chicago Press, 1988), 56-59.

3. Bruce Scott, "National Strategies: Key to International Competition," in *U.S. Competitiveness in the World Economy*, Bruce Scott and George Lodge, eds. (New York: Norton, 1987), 75, 93-94.

4. That is, technological capabilities were assumed to be equal everywhere and therefore by themselves did not offer any country comparative advantage over another.

5. Harry G. Johnson, "Optimum Tariffs and Retaliation," *Review of Economic Studies* 21 (1953), 142-153; Richard Caves, "International Trade and Industrial Organization: Problems Solved and Unsolved," *European Economic Review* 28 (August 1985), 382.

6. The original contributions, which appeared in the late 1970's and early 1980's, are too numerous to cite. An early synopsis was attempted by Gene M. Grossman and J. David Richardson, *Strategic Trade Policy: A Survey of Issues and Early Analysis*, Special Papers in International Economics no. 15 (Princeton: International Finance Section, 1985). Other germane references include Henry Kierzowski, ed., *Monopolistic Competition and International Trade* (Oxford: Clarendon Press, 1984); Paul R. Krugman, ed., *Strategic Trade Policy and the New International Economics* (Cambridge: MIT Press, 1986).

7. Richard Harris, *Trade, Industrial Policy, and International Competition* (Toronto: University of Toronto Press, 1985), 35-43.

8. An economic rent occurs when the return from an input is higher than what that input could earn in an alternative capacity. Orthodox economic theory holds that rent is competed away.

9. This assumes that market size is limited and that no technological innovation

would make existing products or processes obsolete. See Pankaj Ghemawat, "Sustainable Advantage," *Harvard Business Review* 64 (September/October 1986), 179-193. See also Alex Jacquemin, *The New Industrial Organization: Market Forces and Strategic Behavior* (Cambridge: MIT Press, 1987). The subject of this article is discussed in the final chapter of the book, pp. 168-182.

10. Alex Jacquemin, *The New Industrial Organization: Market Forces and Strategic Behavior* (Cambridge: MIT Press, 1987), 168-182.

11. Paul Krugman, "Import Protection as Export Promotion," in *Monopolistic Competition and International Trade*, Henry Kierzowski, ed. (Oxford: Clarendon Press, 1984), 187-188.

12. James Brander, "Rationales for Strategic Trade and Industrial Policy," in *Strategic Trade Policy and the New International Economics*, Paul Krugman, ed. (Cambridge: MIT Press, 1986), 26-34.

13. James Brander and Barbara Spencer, "Export Subsidies and International Market Share Rivalry," *Journal of International Economics* 18 (February 1985), 97.

14. The Cournot model suffers some drawbacks insofar as firms choose output levels and allow prices to change as necessary to sell the product. While this is probably appropriate if quantities are set by physical capacity constraints, it is less appropriate if firms choose prices and easily adjust output as required.

15. Heinrich von Stackleberg, *The Theory of Market Economy*, translated by Alan T. Peacock (London: W. Hodge, 1952), 190-204.

16. James Brander and Barbara Spencer, "Export Subsidies and International Market Share Rivalry," *Journal of International Economics* 18 (February 1985), 85.

17. This conclusion by Brander and Spencer implies that the acting government, in the view of the foreign firm, is credibly committed to its intervention. Thus, the foreign firm must believe that the subsidization would continue even in the event that the attempted profit shifting failed because the foreign firm refused to retreat. Brander and Spencer suggest that government intervention is credible in this sense as governments have an incentive to maintain a reputation for credibility. See James Brander and Barbara Spencer, "Export Subsidies and International Market Share Rivalry," *Journal of International Economics* 18 (February 1985), 84. Alex Jacquemin suggests that government intervention has credibility based on its reputation and/or resources or because of the expected inertia of policies, once adopted. See Alex Jacquemin, *The New Industrial Organization: Market Forces and Strategic Behavior* (Cambridge: MIT Press, 1987), 172.

18. James Brander and Barbara Spencer, "Export Subsidies and International Market Share Rivalry," *Journal of International Economics* 18 (February 1985), 92.

19. James Brander, "Rationales for Strategic Trade and Industrial Policy," in *Strategic Trade Policy and the New International Economics*, Paul Krugman, ed. (Cambridge: MIT Press, 1986), 39.

20. Ibid., 44.

21. Paul Krugman, "Import Protection as Export Promotion," in *Monopolistic Competition and International Trade*, Harry Kierzowski, ed. (Oxford: Clarendon Press, 1984), 180-193.

22. Ibid., 182.

23. James Brander, "Rationales for Strategic Trade and Industrial Policy," in *Strategic Trade Policy and the New International Economics*, Paul Krugman, ed. (Cambridge: MIT Press, 1986), 33.

24. Paul Krugman, "Import Protection as Export Promotion," in *Monopolistic Competition and International Trade*, Harry Kierzowski, ed. (Oxford: Clarendon Press, 1984), 181.

25. Ibid., 186.

26. Paul Krugman, "New Thinking About Trade Policy," in *Strategic Trade Policy and the New International Economics*, Paul Krugman, ed. (Cambridge: MIT Press, 1986), 9.

27. Paul Krugman, "Is Free Trade Passé?" *Economic Perspectives* 1 (Fall 1987), 40.

28. Gene Grossman, "Strategic Export Promotion: A Critique," in *Strategic Trade Policy and the New International Economics*, Paul Krugman, ed. (Cambridge: MIT Press, 1986), 55-57. See also Avinash K. Dixit, "Trade Policy: An Agenda for Research," in *Strategic Trade Policy and the New International Economics*, Paul Krugman, ed. (Cambridge: MIT Press, 1986).

29. Johnathon Eaton and Gene M. Grossman, "Optimal Trade and Industrial Policy Under Oligopoly," *Quarterly Journal of Economics* 101 (May 1986), 383-406.

30. Gene Grossman, "Strategic Export Promotion: A Critique," in *Strategic Trade Policy and the New International Economics*, Paul Krugman, ed. (Cambridge: MIT Press, 1986), 55-57.

31. Johnathan Eaton and Gene Grossman, "Optimal Trade and Industrial Policy Under Oligopoly," *Quarterly Journal of Economics* 101 (May 1986), 402-03.

32. That strategic trade policy benefits the national welfare also presumes that the firm in question is nationally owned. Increased tax revenues owing to potentially higher profits garner only minimal advantages to the government treasury. See Paul Krugman, "Strategic Sectors and International Competition," in *U.S. Trade Policies in a Changing World Economy*, Robert Stern, ed. (Cambridge: MIT Press, 1987), 219.

33. Avinash Dixit and Gene Grossman, "Targeted Export Promotion with Several Oligopolistic Industries," *Journal of International Economics* 21 (November 1986), 233-249.

34. Ibid, 247-248.

35. Richard G. Harris, *Trade, Industrial Policy, and International Competition* (Toronto: University of Toronto Press, 1985), 111-144.

36. David Richardson, "The Political Economy of Strategic Trade Policy," *International Organization* 44 (Winter 1990), 124.

37. Paul Krugman, "Technology and International Competition: Overview," Paper prepared for a National Academy of Engineering Symposium on *Linking Trade and Technology Policies: An International Comparison*, Washington, D.C.: National Academy of Sciences, 1991. See also Keith Bradsher, "High-Tech Industry is Hard to Help With Subsidy," *New York Times*, 2 February 1993, C1.

38. Tyson argues that it has generated a source of high technology job growth and technological spillovers that are not reflected in monetary returns. See Chapter 5 in Laura D'Andrea Tyson, *Who's Bashing Whom: Trade Conflict in High Technology Industries* (Washington, D.C.: Institute for International Economics, 1992).

39. The recent imposition of tariffs on selected European goods in response to European resistance to terminate soybean subsidies suggests that the United States does not hew an absolute free trade line and will respond to other countries' manipulation of trade policies.

40. A recent study by the Institute for International Economics indicated that Japanese trade barriers, transparent and otherwise, cost Japanese consumers between $75 and $110

billion. See "Japan Trade Barriers Lift Costs Sharply," *Wall Street Journal*, 15 December 1994.

41. Paul Krugman, "Market Access and International Competition: A Simulation Study of 16K Random Access Memory," in *Empirical Methods for International Trade*, Robert Feenstra, ed. (Cambridge: MIT Press, 1985).

42. Kenneth Flamm, "Making New Rules: High Tech Trade Friction and the Semiconductor Industry," *The Brookings Review* (Spring 1991), 28.

5

SEMICONDUCTOR INDUSTRIAL POLICY: WESTERN EUROPE AND JAPAN

The semiconductor industry was born in the United States in 1947 with the invention of the transistor by Bell Laboratories. That the industry was once exclusively an American industry was not a matter of chance. The United States had the highest level of R&D as a percentage of GDP, the highest ratio of scientific engineering personnel and capital, and an economy that generated 50% of global output after World War II. However, the subsequent development of the industry in the United States and elsewhere has been closely linked to government policy. In the United States, military considerations represented the principal motivating force for the American semiconductor industry, and defense and space programs provided early R&D contracts and guaranteed markets that accelerated the industry's development. During the early stages of the industry the American military represented 100% of the market for semiconductors—48% of total semiconductor shipments through the mid-1960's—and this provided the economies of scale and capital necessary at a time when semiconductors were prohibitively expensive for commercial use. The military requirements of greater speed and miniaturization also made semiconductors increasingly suitable for commercial applications which became an important market once the development of esoteric technology became more standardized. Military policy had the unintended result of increasing the rate and direction of technological change that helped establish the American industry as the world leader in market share and technological development. However, the importance of this relationship waned as the industry matured and semiconductors found wider applications in the industrial network. By the mid-1970's government procurement constituted less then 14% of overall American semiconductor sales.[1] Although governments in Europe and Japan were not as active at the early stage of the industry's development, the subsequent emergence of government policies in Europe and Asia occurred at the same time as the American government became a marginal factor in semiconductor development. In Japan and Western Europe public policy was shaped by commercial considerations and the recognition that domestic semiconductor capability was desirable, if not necessary, to remain competitive in the growing number of downstream industries in which semiconductors had become important determinants of price and performance.

In the early 1960's military-sponsored development of superior silicon-based semiconductor technology led to the commercial introduction of integrated circuits and secured American market and technological preeminence in the following two decades. Semiconductor competition in these years was fierce between American firms. During this period, many American producers shifted the assembly of semiconductor devices offshore as a cost-cutting move against rival American producers.[2] That the United States began to show a deficit in integrated circuits by 1978 reflected the growth of imports from American-owned offshore production and assembly facilities rather than a loss of competitiveness. Foreign firms remained marginal players in technology and world market share.

The mid-1960's proved critical for the Japanese industry. Technology licensed to Japanese companies in the 1950's permitted the production of germanium-based transistors and through the late 1950's and early 1960's the Japanese industry excelled as a high-volume producer of these cheap commercial devices used in consumer electronics. However, emerging silicon technology rendered Japanese techniques in germanium-based semiconductor devices obsolete. The Japanese were compelled to redirect their efforts toward more sophisticated silicon based technology to remain competitive in light of technology advances. It was at this time that long-term plans were being made that were to change the character of the Japanese semiconductor industry from a high-volume producer of cheap commercial devices to a manufacturer of integrated circuits.

The Japanese government targeted the semiconductor and computer industries by employing a host of promotional measures conceived around the same companies that had become world-class consumer electronic producers. Japanese semiconductor houses had narrowed the substantial technology gap with the United States by the late 1970's and imposed a broad-based challenge to American commercial leadership in the industry by the early 1980's. It is a challenge that has gained in intensity and effectiveness. By the mid 1980's, MITI began to create institutional arrangements that reflected Japan's attainment of technological parity and objectives to extend the frontier.[3]

The full extent of the Japanese assault on the U.S. industry was most evident by 1985. By the end of that year, most American DRAM (Dynamic Random Access Memory) producers were forced out of the market, and American market share of EPROMs (Erasable Program Memory) was curtailed to 50%.[4] American worldwide DRAM market share went from 74% in 1978 to 18% in 1986, while Japanese market share jumped from 22 to 78% during the same period.[5] In overall semiconductor world market shares, Japan climbed from 33%

in 1982 to 51.2% in 1988, while the United States share fell from 56.7 to 38%.[6] Through the 1980's Japanese producers secured a growing percentage of the American and global market in all the major semiconductor subsectors. At mid-decade, Japan held an enormous market lead in DRAMs, had converging markets shares in microcomponents (processors, controllers, peripherals) and ASICs (application specific integrated circuits), and a growing global lead in overall semiconductor market share.

Efforts by Western European governments to create a competitive semiconductor industry remained focused on the national level. Promotion included home-market protection and subsidization of captive producers as national champions. These national strategies guaranteed, at great cost, that Europe would at least maintain an industrial presence in the electronic systems and component industry. However, by the early 1980's it was clear that efforts to arrest the decline in microelectronics were unequal to the task of averting further marginalization. From 1975 to 1985 world market share for European producers fell from 16 to 10%.[7] Japanese inroads into DRAM and EPROM markets, and persistent losses by European firms to American and later Japanese competitors, motivated a drive to find Europe-wide solutions.

The American response to the new competitive conditions was the creation of SEMATECH and the negotiation of the Semiconductor Trade Agreement (STA). SEMATECH, founded in 1987, is an ongoing industry-government R&D consortium designed to shore up competitiveness in the semiconductor manufacturing equipment industry (SM&E's). In the trade arena, American semiconductor producers filed antidumping actions against Japanese firms, and pursuant to Section 301 of the Trade Act of 1974, filed injury claims caused by alleged denial of access to the Japanese market. These legal actions set in motion a string of negotiations between the governments of the United States and Japan. The resulting Semiconductor Trade Agreement was conceived to terminate Japanese DRAM dumping and open up the Japanese market to American semiconductor companies.[8]

The structure of the semiconductor industry in Europe and Japan is different from the American industry. The Japanese industry is dominated by six captive producers that are large, vertically integrated electronics firms that incorporate a significant portion of their semiconductor products for in-house use. With the exception of a few large vertically integrated electronic systems producers (Siemens, SGS/Thomson, Philips), most European semiconductor producers compete against Japan's integrated giants without the benefit of leading-edge captive-chip production. Although several large, vertically integrated captive producers such as IBM and AT&T existed in the United States, their semiconductor production was principally for in-house use and had minimal open-market sales. The American industry has been dominated by numerous but smaller merchant firms that produce semiconductors for sale in regional and global markets. These smaller firms have demonstrated greater speed and flexibility in generating sophisticated custom-tailored innovations, but the greater size and diversified revenue streams of captive producers have made them more resilient in a market prone to cyclical product demand.[9]

Global competition in the semiconductor industry in the 1990's has become more broadly scattered. Korea and Taiwan have joined the competition with enormous government subsidies and heavy front-end investment in semiconductor infrastructure.[10] In Europe, extensive national and pan-European technology programs have been adopted to stabilize the declining fortunes of the European industry, and the American industry has recovered from its nadir in the late 1980's. The rapid advances by Japan in technology capabilities and market share during the 1980's instilled a fear of technological dependency among competitors in semiconductors and related industries. This precipitated a greater government-industry collaboration in the United States and Western Europe, one designed to secure national advantage in the global semiconductor industry.

WESTERN EUROPE

France

The French have generally held that government has a leading role in the promotion of industry. It is an orientation that may be attributed to the fact that France has been centralized longer than other European countries. The government has long regarded a high technology establishment as fundamental for national security, and France continues to employ industrial policy as a means to achieve goals of national security, independence (from the American-dominated Western Alliance structure), and prestige. This has motivated the subsidies, protection, and heavy public investment of R&D in industries such as aerospace, electronics, and nuclear energy. Under the Socialist government of Francois Mitterrand elected in 1981 the view that high technology is key to the future of France has been given greater clarity. Promotion of high technology has evolved from the emphasis on prototype construction of specific high technology products (such as Concorde, Airbus, and the Train Grande Vitesse), to a concern with R&D on a wider front.

The Socialist government launched a microelectronics plan in 1981 with the goal of strengthening the nation's microelectronic sectors through nationalizations and huge injections of state investment funds.[11] An important objective was to secure the French semiconductor market for domestic producers by 1986. The large electronic firms of CHI-Honeywell Bull, Thomson, and Matra were all nationalized and the government committed itself to underwriting the development of an autonomous semiconductor industry. Indeed operating losses and the cost associated with the nationalizations were secondary to the attainment of a national semiconductor capacity.[12] In addition to covering operating losses, from 1981 to 1986 the government committed $820 million for

R&D into leading large-scale integration technology, and an additional $300 million in investment subsidies. Accordingly, the plan encouraged French-only strategic alliances.

The microelectronics plan did not fulfill the goals established in 1981.[13] Despite R&D subsidies, coverage of operating losses for computer firms, and nationalizations, the government failed to achieve national autonomy in semiconductors, and remained behind the state-of-the-art integrated circuits produced in Japan and the United States. The failure of the Plan was a turning point for French industrial policy and how that policy related to the wider European Community. Persistent failure in an area deemed critical by the government occasioned a reassessment of the methods of strategic promotion. It became clear to the French, and by extension to the remainder of the EC, that an EC-wide approach would be necessary to rationalize R&D costs and realize economies of scale. While French industrial policy in semiconductors and general information technology has assumed a more pan-European orientation, the government still promotes the longstanding concept of national champions.[14] The policy to absorb the operating losses of Groupe Bull and SGS-Thomson underlie this expensive strategy.[15]

United Kingdom

The United Kingdom has been less motivated than France in the pursuit of an autonomous capability in selected high technology industries. The British semiconductor/microelectronic enterprise has never been a vibrant component of the national economy. Nevertheless, its central importance was recognized and a general consensus that the British microelectronic industry was particularly weak did encourage the government to adopt a host of remedial measures to support the domestic industry. Through the 1970's Conservative and Labor governments fostered ICL (International Computers Ltd.) as a national champion. Government provided annual subsidies and gave it preferential treatment in national procurement procedures. Government programs first conceived under the previous Labor government gained a greater cohesion in the first years of the Thatcher government.[16] In addition, the government founded INMOS, a state-owned producer of microprocessors and DRAMs, and the Conservative government of Margaret Thatcher implemented a National Strategy for Information Technology under a newly created Ministry for Information Technology.

The first British government-sponsored R&D collaboration was the Joint Optoelectronics Research Scheme begun in 1982 with a budget of $40 million, but the largest and most ambitious government project was the Alvey program around which other promotional programs were conceived.[17] The Alvey program (government funded $350 million over a period of eight years) was designed to bring together the efforts of University laboratories and R&D centers of major firms such as Plessey, ICL, and British Telecom. The program achieved advances in very large-scale integration (VLSI) and computer aided design (CAD) technology, but the British industry was never able to capture the commercial windfall the program was intended to create. Brian Oakley, director of the Alvey project, submitted that lack of reliable long-term capital compromised the potential for successful commercialization by industry.[18]

The results of microelectronic promotion in the United Kingdom have not been successful. Consequently, the Conservative government did not create a successor to the Alvey project and has elected not to support large-scale microelectronic research and development. In 1984 the government sold INMOS to the electronics firm Thorn (after pumping in millions of pounds sterling in loan guarantees), and in 1990 permitted the sale of ICL, the nation's largest computer firm, to the Japanese firm Fujitsu. The Japanese acquisition of the computer giant meant that ICL could no longer participate in pan-European information technology projects. Taken together, the United Kingdom will neither be a significant player in the European semiconductor business nor an active participant in Community-wide initiatives in microelectronics.

Germany

Promotion of national champions and other tools of sectoral financial support has been used more sparingly in Germany than in either France or Great Britain. The government has provided aid to various industries such as aerospace, semiconductors, and computers, but has not engaged in the process of picking sectors to support to the same extent as have France and Great Britain. Government policy has generally been to improve the general environment in which the private sector operates rather than provide aid for specific industries. Economic adjustments are, as successive German governments have declared, largely a private and not a public responsibility.[19] While Germany rejects sectoral promotional policies and the national champion approach in principle, the government has implemented a coordinated industrial policy for microelectronics since 1974, has supported R&D, and has subsidized investments of private firms in the semiconductor industry.[20]

The Ministry of Research and Technology (BMFT) has been the conduit of government aid since its creation in 1972, and is concerned solely with the support of high technology industries and the commercial applications of applied research. Germany had the first program in Europe for the direct support of semiconductors between 1974 and 1978, in which the German government provided R&D support to promote the development of an efficient domestic components industry. However, with the continued poor performance of German producers in the integrated circuit market, the government changed its emphasis from basic and applied research to financing more projects for the application of new products into the market. From 1979 to 1983, R&D support and investment subsidies in the semiconductor industry averaged $60 million annually. In 1984, the government launched a more ambitious four-year information technology program[21] that invested almost $350 million in semiconductor R&D.

The goal of becoming a world leader in information technology and semiconductor autonomy has been reflected by an ever-increasing financial commitment. The latest four-year (1990-1993) R&D budget for the BMFT was $2.75 billion.[22] Roughly one-half of this sum was to be dispensed on R&D projects conducted by the five leading industrial concerns (Siemens, Telefunken, AGG, Alcatel/SEL, and Philips). In addition, the BMFT supports the Association of Applied Research Institutes (FHG), which contracts work out to semiconductor producers and coordinates industry-government-university R&D, and has been the principal agent for diffusion of technology to German industry.[23] The government has also harnessed huge R&D projects to develop the next generation of devices through specific programs driven by the quasi-public FHG network.

The motivating force for sectoral intervention in semiconductors has been the fear that domestic electronic producers have become overly dependent on foreign sources of semiconductors. In some circles, however, this is not a cause for alarm. Opponents of German sectoral promotion in semiconductors argue that it is cheaper to buy Japanese and American advanced technology. In addition, the critics argue that the prosperity of the German economy depends less on the success of its high technology industries than it does on the successful diffusion of high technology into the national industrial network. The opposing view regards semiconductor promotion as necessary to avoid potential industrial blackmail by the Japanese.[24] It is this latter view that drives German industrial policy in semiconductors.

The European Community

Since the mid-1970's, there has been growing anxiety in Europe over the relative decline in productivity and real growth rates. At the core of this concern has been the widening gap in the trade of technology-intensive goods.[25] In December 1975 the Economic Outlook of the OECD stated:

> Despite considerable government interest in and encouragement of high technology in Europe over the years, the secular decline of the trade balance in high technology goods continued. This could imply that in the long run Europe will become a special kind of low development area, supplying the rest of the world with food, raw materials and low technology goods.[26]

National policies in Europe through the 1970's and 1980's did little to deter American and Japanese firms from gaining in high technology electronic markets that were once the privileged domain of national champions or local producers. Coincident with this was a steady decline in Europe's share of world semiconductor consumption from 28% in 1970 to 17% in 1984.[27] Even with the relative decline in the consumption of semiconductors, the trade deficit in integrated circuits actually continued to rise, reflecting a growing inability to provide for even decreasing demand. A steady erosion in high technology markets and low economic growth relative to Japan and the United States fostered what came to be known as "Europessimism."

National champion policies were designed to create a few large, vertically integrated electronic producers that were competitive in final markets. Scant attention was given to promoting a merchant semiconductor industry outside the framework of national champions. European semiconductor promotion was, therefore, closely linked to the promotion of computers. It was believed that a large and growing computer industry would in turn promote a vibrant semiconductor industry. However, when the rapid advances in semiconductor technology by the United States in the 1960's led to wider and more efficient applications in the computer industry, the relatively weak European computer industry could not transmit a significant pull on the semiconductor industry. Rather, American computer producers continued to dominate the European computer market with an 81% market share as late as 1983. By contrast, the United States computer industry was the largest in the world and exerted a strong demand for high-performance computer chips, and this further encouraged technological development in the highly competitive United States semiconductor industry. A growing demand for advanced semiconductors was not captured by European producers. State-of-the-art devices were purchased from American firms or from the subsidiaries of American companies in Europe. Consequently, a growing demand for large-scale integrated devices did not result in the entry of European merchant producers into the market and instead marked the steady decline of European semiconductor performance.

Significant technological advances in integrated circuits occurred in the 1970's with the introduction of the microprocessor by Intel in 1971 and with the growing sophistication of memory devices. Semiconductor devices that improved established integrated circuit technology with between 100 to 100,000 circuit elements came to be known as Large Scale Integration (LSI).[28] It enlarged the range of applications for semiconductor technology beyond the simple logic functions of earlier integrated circuits to system and subsystem functions, and led to a greater incorporation of semiconductor devices in end-market products. For example, microprocessors and solid-state memory devices formed the core of new computing systems, ones that enabled the development of new computer concepts such as the minicomputer and the personal computer. Large-scale integration (and later very large-scale integration) capabilities increased the degree of dependence on semiconductor devices and the interdependencies among computers, electronic consumer goods, and telecommunications equipment illustrated the convergence of demand in the electronics final markets.

Through the first half of the 1970's, the European semiconductor industry continued to produce discrete devices, linear integrated circuits, and specialty and custom-integrated circuits, but did not commit itself to competing with the American firms in the LSI market. By the second half of the 1970's, European producers recognized the declining utility of these devices and the increasing importance of LSI in overall semiconductor consumption. They also became aware of the threat that a continued inability to produce LSI devices posed as the convergence of electronic end markets using those devices threatened the established competitiveness of the vertically integrated producers in consumer electronics, computers, and telecommunications equipment. European governments concluded that a domestic LSI integrated circuit capability was a strategic necessity for reasons of national security and for sustaining linked electronic and manufacturing industries as a whole. It was not until the late 1970's that specific measures were undertaken to strengthen the semiconductor industry directly.

It was in this context that the microelectronic plan under French President Mitterrand promoted the development of commodity memory chips. Prior to 1978, French and British governments had channeled funds through the domestic computer industry without a coherent plan for the development of an integrated circuit industry. Instead of implementing a vigorous technological effort in advanced memory semiconductors, French and other European industrial policies had maintained general support for the electronic champions, which in turn, found it more profitable to concentrate on products for telecommunications and related industrial markets where European producers remained strong.[29] Moreover, national champions were primarily interested in producing semiconductors for in-house needs and had only secondary interest in producing commodity components for the smaller electronic producers who constituted the bulk of the electronic sector.[30]

It took time before the new large-scale integration technology and inter-industry linkages generated by that technology would have their full effect on the European semiconductor industry. Government and industry in Europe became committed to large-scale integration technology but remained behind the technological frontier and continued to see its share in of world-wide semiconductor production fall. In order to improve their position in the market, many European producers purchased American LSI merchant producers, and as indicated in the next section, began R&D collaboration with other semiconductor producers. However, late entry, the complexity of LSI technology, and the cumulativeness of technological advance constituted major disadvantages for European firms because they lacked the technological capabilities and experience. Unlike the semiconductor technology during the transistor period (1950's), LSI technology was complex, and advanced technological capabilities were required to transmit the technology or to develop the new technology.[31] Despite the commitment of company resources to production and R&D, and the support given by governments, European producers were not able to reach the technological frontier or to increase their market share in standard LSI device markets, and this in turn compromised their ability to compete in next-generation VLSI markets. While European producers were finding their way to large-scale integration technology, the American and Japanese industry had already begun to market next-generation, very large-scale integrated circuit semiconductor technology.

European government policies were aimed at reducing the technological lags of domestic producers and at developing a productive capability in the most advanced technology areas. It was hoped that with the promotion of national champions and the maintenance of high tariffs, European companies could challenge the market dominance of electronic giants such as IBM. However, protection of the semiconductor and computer industries with high tariffs (17% ad valorem, 14% after 1985) created a policy conflict. A higher cost for imported semiconductors led to higher prices and diminished sales for European computer makers. A lower semiconductor tariff would lower the price input in computers but would jeopardize the local semiconductor industry. This dilemma was resolved by maintaining support for both industries while allowing foreign direct investment in Europe by foreign semiconductor producers. Although this led to a large foreign chip-production base inside Europe, it did not lead to a transfer of technology to European producers. The European computer industry benefited from cheaper access to state-of-the-art semiconductor technology, but this was proprietary technology held by American chip manufacturers.

State-subsidized projects and policies failed throughout Europe to create a competitive semiconductor industry. Nationalistic strategies encouraged an inward orientation. High tariffs on semiconductors provided a degree of insulation from international competition, while government subsidy and procurement preferences protected national champions from domestic competition. With little European inter-country competition and a high proportion of national demand reserved through state policy for those firms, there was little incentive for technological innovation. Protection did not promote efficiency. Instead it insulated huge national firms from the discipline of competition and made them complacent in their protected home markets. Lack of competition negated the putative benefits of economies of scale that had motivated the formation of the large national champions thought necessary to compete with Japan and the United States.

Compounding these weaknesses was the fragmentation of the European semiconductor market. Fragmented among national markets, European companies could not achieve the scale economies in R&D, production, and marketing needed to cover large front-end investment costs. This eliminated the potential for Europe-wide product specialization. Moreover, because national policies discouraged cooperation with rival European producers and encouraged cooperation with American producers (because of required American technology), it was American producers who captured the economies of scale denied to European producers.[32] European firms remained at a disadvantage and were consigned to a second-class status.

Changing Conditions

Local producers found it increasingly difficult to meet the standards of international competition in quality and price. By the mid 1980's, public policy toward semiconductors had to change in order to prevent these sectors from becoming permanent wards of the state. The state of the industry was such that European firms were not in and may never be in the position to challenge the international market position of Japanese and American firms. Limited returns despite expensive promotion suggest that what was once considered a strategic necessity may have become only a misallocation of national resources. In this regard, one might presume that the marginalization of the European industry by the United States and Japan should be accepted as a fact of life, and largely unimportant, provided the supply remained secure and competitively priced. However, being uncompetitive in semiconductors does present a potentially dangerous problem for Europe. The United States and Japan are dominant chip producers and Europe is largely a chip consumer as European producers account for under 40% of European semiconductor consumption. Two strategic threats inhere to such a dependent position: one is that producers may restrict supply and realize supernormal profits,[33] and the second is that higher chip prices can make them less competitive than vertically integrated chip suppliers who are also competitors in downstream markets. If chip suppliers are vertically integrated, then they can use a technological advantage by integrating that advantage forward into systems industries. It is conceivable, therefore, that costly domestic semiconductor promotion may be less expensive to local users if a dominant position in technology for foreign suppliers allows them to extract monopoly rents or, perhaps worse, withhold the delivery of vital components.

European electronic systems producers have relied heavily on American components suppliers. These American producers were numerous merchant producers located in Europe and the United States, and not in direct competition with the European systems producers they were supplying. Despite an overreliance on American components, intense technological competition among American merchant semiconductor producers provided a reliable flow of state-of-the-art products at reasonable prices. However, the ascendancy of the Japanese semiconductor industry in world markets in the early 1980's jeopardized that supply trajectory. A potential supply problem emerged because Japanese chip producers were also large vertically integrated electronic systems producers that competed in the same downstream electronic systems markets with their semiconductor customers. The possibility arose that a competing electronic systems producer could withhold timely access to a vital input in order to gain advantage in downstream markets.

With Japanese producers gaining in the European market, the security of supply became less certain. This was most evident during the DRAM shortage of 1987 to 1989. There is evidence Japanese companies, after securing close to 90% of the world market in state-of-the-art 256K DRAMs, acted in concert with

MITI to reduce production to realize artificially high prices on the world market.[34] In addition, many Japanese DRAM suppliers preconditioned the sale of DRAMs to firms by tying in supply contracts for other components.[35] The withholding of critical components and the subsequent DRAM shortage highlighted certain risks associated with excessive dependency on foreign suppliers for critical electronic components. In the United States, an investigation by the General Accounting Office revealed threats of strategic withholding to elicit the license of technology and or the advance commitment to purchase chips.[36] This has troubling implications for timely and cost-effective access to other important high technology products. Thomson Consumer Electronics, a French company that competes with Japan in VCRs, indicated that "the Japanese have stopped delivering the most recent components we need to manufacture new tape recorders."[37] The General Accounting Office reported cases of American electronics firms being damaged and in some cases driven to bankruptcy for being refused state-of-the-art displays, an industry dominated by the Japanese.[38] This strategic withholding illustrated the danger of overreliance for a cost-sensitive input, and it created a new menu of concerns for corporate and public policy.

The growing support for semiconductors on the national and regional level in the mid-1980's reflected these mounting concerns, and defense against the strategic control of a vital input meant strengthening the European base in semiconductor technology. During the 1980's European national governments made concerted efforts to promote their semiconductor industries through national policies, and after 1984, through regional initiatives. As discussed earlier, past national industrial policies tended to overlook the merits of regional cooperation in high technology.[39] Cognizant of the limits associated with a nationalistic technology policy, national governments were increasingly won over to the notion of a European industrial policy. Domestic insulation, R&D subsidies, and international tariffs had done little to foster a competitive market presence, and within Europe it became clear that European companies had to set sights on global competition to compete successfully, even in Europe.

An important element in the success of any European industrial policy for semiconductors is the realization of a single common market that ensures the free movement of goods, services, labor, and capital among member countries. This should provide economies of scale and efficiencies that enable high technology firms to compete more effectively on open markets. The Single European Act has gradually centralized European trade authority in the EC Commission. And, despite assurances that movement toward a united Europe 1992 will not lead to a fortress Europe, recent standards and rules have established a preference and advantage for European semiconductor products.

The most important of these measures has been a 1989 EC Commission regulation changing the test for rules of origin to favor components manufactured in Europe.[40] Previously, the EC (like the United States) defined rules of origin as the country where the last substantial process or operation that was economically justified was performed. Under this definition, Japanese and American semiconductors manufactured outside the EC but assembled and tested in Europe would be considered of European origin and eligible for national procurement programs[41] and exempt from tariff and antidumping regulations. The new definition hardens the test by insisting that a core manufacturing process be conducted within the EC. The only way to establish European origin now is to establish full-scale manufacturing operations within the Europe. The growing pressure to source European, coupled with an external tariff of 14% on semiconductors, 9% wafers, 5.8% on semiconductor parts, and 4.9% on computers, suggests that the trade procedures being centralized in Brussels bear a distinct protectionist flavor. Not surprisingly, the fear of further protectionist policies has spurred an extensive wave of semiconductor manufacturing investment strategies in Europe by American and Japanese companies eager not to be shut out by Europe 1992.

Reinhard Buscher, a senior official of the European Community's Executive Commission in Brussels, stated: "We have tried to use this old-fashioned expression—industrial policy—and fill it with new content. The old style of subsidies, quotas, insistence on local content and other protective measures is over and finished."[42] The preceding section indicates, however, that many protective measures remain securely in place. Moreover, protection and issues of rules of origin are not the only components of the latest wave of semiconductor promotion. The centerpiece of the European effort is the trans-European cooperative research and development projects in precompetitive technologies. The Single European Act that was formally enacted at the close of 1992 was motivated in part by the poor performance of its leading high technology industries.[43] Competitiveness has assumed a regional focus. Through the elimination of barriers between the member states and the gradual regionalization of national R&D subsidies, Europe hopes to achieve the economies of scale and economic efficiency that will make it more competitive with Japan and the United States.

There are three venues of government-sponsored R&D activity. One tier remains at the national level depicted in the country sections of Germany and France. Another tier, and the principal thrust of European efforts in semiconductors and more general information technology, is in government-sponsored transnational R&D projects. One major group of transnational projects is sponsored by, and under the aegis of, the EC machinery in Brussels. The other major group consists of national government collaboration with large companies beyond the legislative purview of the EC. These have been loosely labeled EUREKA.

Closing the technology gap is a multifaceted task. Cooperation in R&D projects (owing to rising costs and comparative advantages associated with expertise), production (to realize economies of scale), and also in marketing (to gain foothold in foreign markets) are being applied by Community-wide programs. Cooperation has been founded on a distinctly European consensus to promote perceived European interests. It is an orientation that recognizes the problem that European firms, lulled by the prospect of more advanced technology and markets, may be more inclined to participate in partnerships with American and Japanese concerns. Although this may work well for the individual firm, in the aggregate it could be harmful to European high technology competitiveness if the relative weakness of European technology consigned European participants to the role of subcontractors. In fact, the immediate motivating factor behind transnational R&D consortiums was to respond to the American Strategic Defense Initiative and the fear that European technology would be consigned to greater obsolescence. Community-wide initiatives to strengthen the European R&D network provide an alternative and distinctly technological development trajectory.

ESPRIT (European Strategic Program on Information Technology) was created in 1984 and is a program of precompetitive research projects with European enterprises, universities, and research centers. Projects are financed 50% by EC funds, the remainder coming from industry participants. With an R&D budget over $1.5 billion for the first five years the program covers microelectronics, software technology, office automation, computer-integrated production, and advanced information processing.[44] The goals of the program are to encourage European upstream industrial cooperation and to avail European enterprises of the technology that will contribute to their market position over the following 10-year period. It facilitates the linkages between enterprises and member countries and stimulates cooperation among private enterprise, universities, and research centers.[45] The program operates in two five-year cycles and has the goal of achieving parity with America and Japan in semiconductor and general information technology. (ESPRIT II (1989-1994) was twice as large as ESPRIT I (1984-1989) It has served as a model for other trans-European high technology projects.[46]

However, ESPRIT has attracted criticism. Many larger European companies disagreed with a policy of helping less-developed EC member states catch up, as opposed to enabling the more advanced EC states to overtake Japan and the United States.[47] These concerns were reflected in the initiation of large joint R&D projects among the most advanced EC member states. In 1984, Germany and the Netherlands pledged financial support for the Megaproject, a Siemens-Philips effort to develop the 1M SRAM and the 4M DRAM. In the following year, French President Mitterrand proposed EUREKA (European Research Coordination Agency). EUREKA was conceived to promote the development of products with high growth potential in world markets and ones that require cooperation in resources and expertise (supercomputers, high-powered lasers and particle beams, artificial intelligence, and high-speed microelectronics, among others). Its approach is market driven as private enterprise assumes the initiative and raise requisite funds for the investments. Projects range from precompetitive R&D to production and marketing but generally focus on the advanced development and commercialization stage of technology while EC led efforts generally commit resources to basic research themes. The primary function of the agency is one of a broker, assimilator of information of ideas, and consultant between potential partners while stimulating project development. Funding for EC programs comes out of the EC treasury while EUREKA funding derives half its revenues from member governments and half from project participants.[48] The total financing cost of projects agreed under EUREKA through 1990 amounted to 7.6 billion ECU.[49]

The EUREKA program consists of 18 partner European countries (the EC being the 19th member). EUREKA is basically a label placed on joint market-oriented R&D projects in the advanced technologies involving more than one member state. Participation by non-European members has been a thorny issue. In principle, EC subsidiaries of American companies with European research facilities are eligible to participate in EUREKA projects. However, the inclusion of IBM Europe in the EUREKA program for semiconductors, JESSI, was initially contingent on access to the American SEMATECH consortium which excluded all foreign participation.[50] It appears, however, that subsidiaries of American and Japanese concerns may eventually gain a wider entry provided that the decision centers of those firms are fundamentally European. The stimulation of European industry may benefit from such an association. However, as the EUREKA declaration of principles suggest, it is a movement to revitalize European industry:

> The objective of Eureka is to raise, through closer cooperation among enterprises and research institutes in the field of advanced technologies, the productivity and competitiveness of Europe's industries and national economies on the world market, and hence strengthen the basis for lasting prosperity and employment: Eureka will enable Europe to master and exploit the technologies that are important for its future, and to build up its capability in crucial areas.[51]

The Joint European Submicron Silicon Initiative (JESSI) is a government-industry R&D project devoted exclusively to semiconductor technology. It was conceived as an offshoot to the Philips-Siemens Megaproject which ended in 1989. The Philips-Siemens project was a success as it led to the development of 4M DRAM and 1M SRAM for Siemens and Philips respectively. However, both these firms recognized that they still had a small share of the European market and still depended heavily on Japanese semiconductors and semiconductor manufacturing equipment and materials. Accordingly, these two firms proposed a follow-up effort, and in 1989 JESSI expanded and received EUREKA designation. The primary goal for the 28 participating companies during the eight-year project is to invigorate the entire microelectronics chain in Europe and enable European firms to produce next-generation 64 megabit devices. The budget of $3.7 billion is funded 50% by member companies, 25% by national governments, and the remainder by the EC.[52]

In addition to developing the next memory chip generation, the new effort will address the broader aspects of competitiveness in semiconductor research, production, and manufacturing technology. The JESSI Planning Committee has outlined four principal areas of concern that include semiconductor manufacturing technology, applications, equipment and materials, and various areas of additional basic research. The JESSI agenda is larger than the American SEMATECH consortium for semiconductor manufacturing technology and any single Japanese collaborative R&D effort.

Results of Promotion

The semiconductor promotion results under the first five years of ESPRIT have been mixed. There were "valuable" technological advances in computer-aided design, and projects in silicon technologies "had been successful in so far as the widening of the technological gap had been arrested."[53] Viewed against the broader goal of technological parity with Japan and the United States however, the first five years fell well short of its projected ten-year goal. The monetary investment in ESPRIT has been large and the technological results "unimpressive"[54] given the scale of the effort. The ESPRIT Review Board stated that many projects lacked strategic direction, something corrected in the second phase of ESPRIT (ESPRIT II) as it has become more focused in its research and in coordinating related activities with JESSI.

Some observers have criticized ESPRIT for excluding non-EC member companies. Partnerships with American and Japanese companies may be necessary to gain access to advanced manufacturing technologies that European companies do not have. A potential problem is that EC-driven financial support may promote inefficient or suboptimal partnerships. For example, in 1983 Siemens commenced a drive to catchup to Japanese DRAM technology by 1990. The creation of the Megaproject was one aspect of this campaign. Another aspect, however, was a deal with Toshiba in which Toshiba transferred its design and production know-how for 1M DRAMS to Siemens, a deal that helped Siemens in its drive to produce next-generation 4M DRAM in the Megaproject. In this case, access to technology not available in European companies was integral to the technology goals of the Megaproject. Some critics argue that ESPRIT and JESSI are weakened by the growing interdependence of members in other types of alliances such as the one indicated above. Others contend however, that technological advances are compromised by a political xenophobia that is counterproductive commercially.[55] The tension between promoting pan-European partnerships and company-level best practices is yet to be resolved.

Although it is too early to judge the results of JESSI, the credibility of the research effort has suffered in the past several years. The Dutch electronics firm Philips N.V. was forced to scale back its commitment owing to financial difficulties; ICL plc of Great Britain was expelled from key projects after its acquisition by the Fujitsu Corporation of Japan; and Siemens of Germany teamed up with IBM to develop new generations of memory chips including the

JESSI-sponsored 64 megabit DRAMs. Furthermore, budget cuts in the EC have compromised the full financial commitment of the EC, and slow payments by national governments during 1991 forced JESSI to cut its budget by 25%.[56]

The national promotional policies of European governments in the 1970's through mid-1980's failed to create a technologically advanced and competitive European semiconductor industry. Home market protection and the promotion of national champions in semiconductors have been expensive and have not led to an autonomous capability. The case of France illustrates that massive government funding has been required to ensure the survival of large semiconductor producers. The government was obliged to underwrite the operating losses at Thomson and Bull, which amounted to almost 6 billion Francs in 1984-1985.[57] However, the Industry Minister of France stated that "policy at Bull has not been an error because France still has a computer industry which is vital."[58] While such a statement may be politically motivated, huge infusions of government funds have enabled the growth, technological advance, and survival of semiconductor producers that probably would not have survived pure free market competition. However, it is questionable that this is an enviable position. Government aid has enabled the largest French electronics producers to remain viable players in the industry, but at great cost to French taxpayers. The continued inability to translate government support into competitive and profitable semiconductor producers suggests that such government expenditures more likely represent a misallocation of national resources.

The important difference to semiconductor promotion in Europe in the mid-1980's and after has been the gradual Europeanization of national semiconductor policies. National subsidies remain but they are applied in a different competitive environment. The Single Market and the emergence of Japan as a leading semiconductor producer have sharpened the focus on semiconductors and have fostered a greater degree of competitive discipline. Massive funding in transnational industry-led projects is the centerpiece of the new European policy for semiconductor promotion. The pooling of continental resources and procompetitive changes in the semiconductor industry suggest that semiconductor promotion in the latter 1980's and after is quite different.

An optimistic interpretation of the new model is that these initiatives are going to provide the European semiconductor industry with a competitive position in semiconductor technology and markets. A more pessimistic interpretation is that European governments are sinking enormous funds into an industry that continues to lose ground to Japanese and American competitors. The European semiconductor industry remains a distant third in the competitive race.[59] As Figure 5.1 indicates, efforts to expand European semiconductor market share in Europe have not succeeded. Europe continues to be dominated by foreign producers. The poor results of the preregionalization period of semiconductor promotion are illustrated in Figure 5.2. From 1975 to 1985 European semiconductor manufacturers steadily lost semiconductor world market share. However, since 1986 the European world market share has not rebounded and

Figure 5.1
Semiconductor Sales in Western Europe, 1980-1993

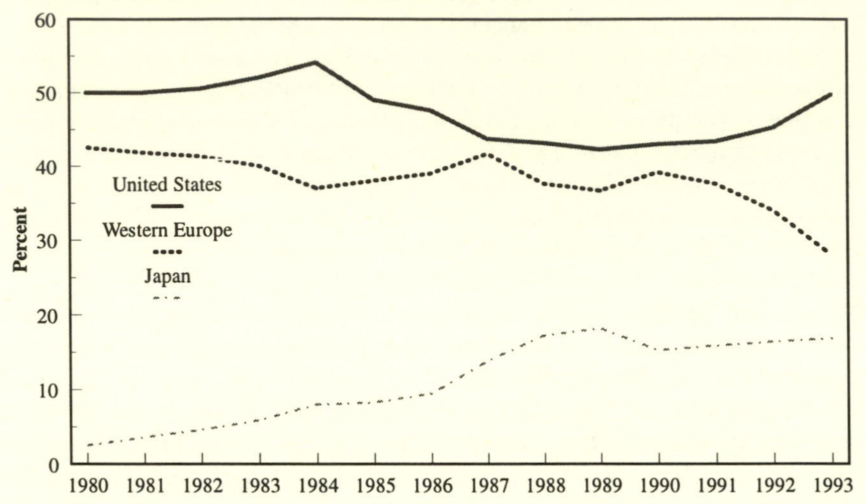

Source: Dataquest

Figure 5.2
European Share of the World Semiconductor Market, 1970-1993

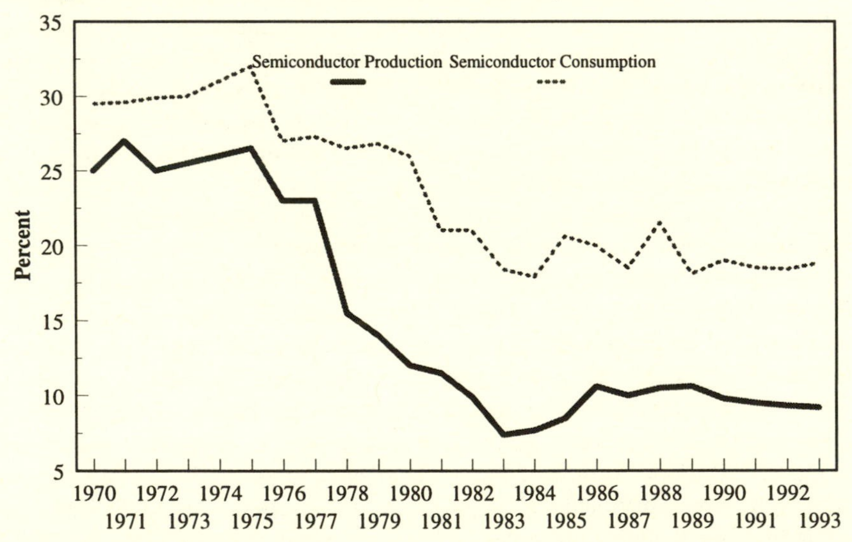

Source: Dataquest

has instead fluctuated from 10 to 12%, stabilizing more recently around 11%. Europe remains the third largest semiconductor market in the world ($11 billion semiconductors annually), behind Japan and the United States. It is too early to measure the ultimate success or failure of transnational semiconductor promotion in ESPRIT and JESSI. JESSI is viewed as the best hope for the European semiconductor industry.[60] At best, the future may yield technological advances that provide a cost-effective contribution to the European semiconductor infrastructure, and at worst, it may divert resources away from other promising alternatives and prove to be a waste of money. However, these transnational initiatives have already succeeded in bringing the main semiconductor houses together to form linkages, networks, and groupings that accelerate the exchange of information and ideas. For example, there have been a number of strategic partnerships in memory development since 1983 among Europe's leading semiconductor manufacturers.[61] Government funding in the Philips-Siemens Megaproject has resulted in memory chips of technological capability approaching that of leading American and Japanese companies.[62] Yet while government promotion of semiconductors on a national and supranational basis has enhanced the level of semiconductor technology, on balance this has not yet appeared to have provided any broad-based and discernible increase in global or local market share for European producers.

JAPAN

Government promotion and industry protection have had more favorable results for the Japanese semiconductor industry. Since the mid-1980's Japan has challenged the United States as the world leader in semiconductor market share, in research and capital investment levels, and in the development of device, equipment technologies, and materials. The semiconductor industry was created and dominated by the United States, but American firms have lost the commanding lead they enjoyed in every area of the industry, and while there is some disagreement concerning the overall impact of Japanese government promotion, the spirit and prevalence of government intervention in the industry has been undisputed.

The Ministry of International Trade and Industry (MITI) has been the principal purveyor of Japanese industrial policy in the semiconductor industry. A main objective of semiconductor promotion was to make the Japanese computer industry more competitive. From the start, semiconductors were considered a vital link in the computer network, and it was hoped that advances in the Japanese semiconductor industry would drive advances and success in the computer industry. Japanese electronic firms began manufacturing computers in the late 1950's and relied on government policies to narrow the technology gap. MITI regulated virtually every aspect of the computer market and government procurement was subject to a formal "buy Japan" Policy.[63] Nevertheless, throughout the 1960's IBM was able to command 50% of the Japanese market, compared to 5% for the new Japanese computer industry.[64]

Infant-industry protection was a mainstay of Japanese promotion. Until 1976, high tariffs, restrictive quotas, and investment restrictions protected the domestic market from American imports. This protection accelerated development of the local semiconductor industry in several crucial ways. First, limited foreign access allowed domestic firms to mature by providing a built-in market for Japanese products. Although the United States occupied half the Japanese market in computers, its share in the semiconductor market remained very small. In 1975 American firms had 98% of the American semiconductor market, 78% of the European market, but under 20% of the Japanese market.[65] Second, in response to limited opportunities in the Japanese market, many American semiconductor firms licensed their technology to Japan as a second-best alternative. While the sale of advanced technology to Japan was an easy way to turn a quick profit, it also facilitated Japanese technology development. According to the National Research Council, Japanese companies paid $17 billion for the transfer and adaptation of foreign technology from the 1950's to 1970's, and although royalty fees were burdensome for some Japanese companies, the benefits outweighed the marginal costs since those fees represented a fraction of what it would have otherwise cost to develop.[66] The technology transfer saved time, clarified their R&D agenda, and freed resources to focus on incremental adaptations. This enabled the Japanese industry to advance rapidly across many fronts. Yet because the United States was the largest market for semiconductors and dominated every important segment of the industry through the 1970's, it was not obvious that this technology transfer was short-sighted.

Investment restrictions in Japan during those years were extensive. Foreign ownership in wholly owned subsidiaries, participation in joint ventures, foreign purchases of equity in Japanese firms thwarted by MITI, and foreign investment in semiconductors were subject to government approval, which was very rarely given.[67] In effect, the Japanese government shut off its domestic market to foreign competition. A report commissioned by the Office of the U.S. Trade Representative concluded that the American share of the Japanese semiconductor market in the 1980's was half of what it would have been had American firms been able to invest in production and marketing facilities in Japan during the 1960's and 1970's.[68] These restrictions also applied to electronic systems producers in telecommunications, computer, and consumer electronic industries. In some measure, American semiconductor producers have not been able to gain a greater share of the Japanese market precisely because the electronic systems producers that constitute the demand base for semiconductors have also not been able to penetrate the Japanese market.

Foreign pressure, most notably from the United States, forced Japan to liberalize its trade regime by 1976. One important result of this was that the government moved to ensure that microelectronic industries would be able to compete in the new competitive climate.[69] Japan continued to lag behind in the most advanced semiconductor technologies, and at this time support for and coordination of research and development efforts became a major ingredient in the promotion of the semiconductor industry. In 1971 the Law for Provisional Measures to Promote Specific Electronic and Machinery Industries designated three strategic categories to promote: advanced technologies where Japanese firms lagged behind American firms; production technologies demanded in large-scale integrated circuit production (LSI); and high-volume production technologies. The national telecommunications company, Nippon Telephone and Telegraph (NTT), extended financial and procurement support to Japanese semiconductor equipment and material suppliers, encouraged them to forge links with Japanese semiconductor producers, and carried on a large amount of basic research in semiconductor technologies. MITI limited entry into the industry by promoting three paired groups of the six dominant semiconductor firms, funneled soft loans to these firms, and encouraged them to cooperate in precommercial R&D as well as in production and marketing of LSI devices. Between 1972 and 1976 these groups received $600 million in research subsidies.[70] Cooperation in R&D and government policy reduced duplication, brought together firms of different technological capabilities,[71] and reduced the cost of R&D in an ever-increasingly complex technology.

Unlike their European counterparts, Japanese semiconductor producers were committed to LSI technology and developed, perhaps coincidentally, enabling technologies from the start. MOS technology gradually replaced bipolar technology as the most widely used technology in LSI digital devices because MOS integrated circuits consumed less power and had higher density capabilities, properties essential for the development of memory devices and microprocessors. In the late 1960's and early 1970's the primary impetus to develop MOS integrated circuits derived from the market for calculators, a market that accounted for over 16% of total Japanese demand for semiconductors.[72] Further demand for LSI devices came from a growing computer industry. By the mid-1970's the Japanese computer industry had grown from 48 to 58% of the domestic market.[73] And by the time of trade liberalization in 1976, the Japanese industry had enhanced its capabilities in more sophisticated semiconductors. Between 1971 and 1976 the industry had doubled the level of sophisticated semiconductors as a share of total semiconductor output. Japanese producers dominated the domestic market for all but the most advanced integrated circuits.[74]

Public policy, demand factors, and an early commitment by Japanese firms to MOS technology helped the Japanese compete in LSI device technology, and in 1979 Japan recorded its first trade surplus with the United States in integrated circuits. Despite these gains, however, the most advanced memory and logic semiconductors used in computers and telecommunications continued to be produced by the United States, and the Japanese computer industry remained a marginal player in global markets despite a decade of promotion. It was believed that if Japanese computers were to make a challenge in the international markets it would have to do so on the back of the next-generation of semiconductor technology.

In order to meet these needs without relying on American firms, MITI organized the Japanese industry with the goal of reaching technological parity with the United States. The result was the very large-scale integrated circuit project between 1976 and 1979, organized as part of a fourth-generation computer project. The goal of the program was to catch up to the United States in production of advanced integrated circuits. The VLSI project included the five leading semiconductor producers, NTT, and MITI's Electrotechnical Laboratory.[75] The total budget for the four year project (1976-1979) was $325 million, of which half was provided by government funds. The project positioned Japanese firms to capture 60% of the world market for 64K DRAM chips and enabled them to commercialize the next-generation of 256K DRAM chips ahead of American competitors. Figure 5.3 indicates that the Japanese world market share for DRAMs was around 24% in 1978, but with the commercialization of the VLSI results, it shot up to 60% in 1984, peaking at 80% in 1987. Global share of DRAMs for American firms followed the opposite trajectory. The project brought Japanese firms to virtual technological parity with American firms and positioned them to be the world leader in the most sophisticated memory devices in the 1980's.

As a complement to the VLSI research agenda Japanese producers placed emphasis on improving the manufacturing process of semiconductors and were subsequently able to parallel advances in technology with gains in quality, productivity, and reduced cost. Indeed Japanese producers leveraged their entry into the American market by engineering a superior system of commodity component manufacturing. Japanese producers took DRAM production to a new level by investing in large, highly automated facilities that translated into lower unit costs than American DRAMs and with fewer defects. They also invested heavily in production capacity for the 16K DRAM during the semiconductor recession of 1975 and were well positioned to take advantage of a postrecession demand that American firms did not prepare to meet. Similar investments were made during the 1981-1982 chip recession, and for the first time, with the 64K DRAM Japanese firms were the first to produce and market a semiconductor device ahead of the Americans. Superior manufacturing and process skills have subsequently furnished Japanese producers with major advantages in commodity chip production.

The loss of strength in memory markets has been particularly worrisome to the American industry, since leading-edge memory devices such as DRAMs drive technological advances in a broad range of process and manufacturing areas. As "technology drivers" DRAMs generate learning applicable to the manufacture of virtually all other semiconductor device types.[76] Moreover, as the highest-volume product market, DRAMs generate sales revenues fundamental to sustaining long-term investment in next-generation R&D and production capacity. The targeting of DRAMs provided experience in large-scale pro-

Figure 5.3
Worldwide DRAM Market Share, 1978-1993

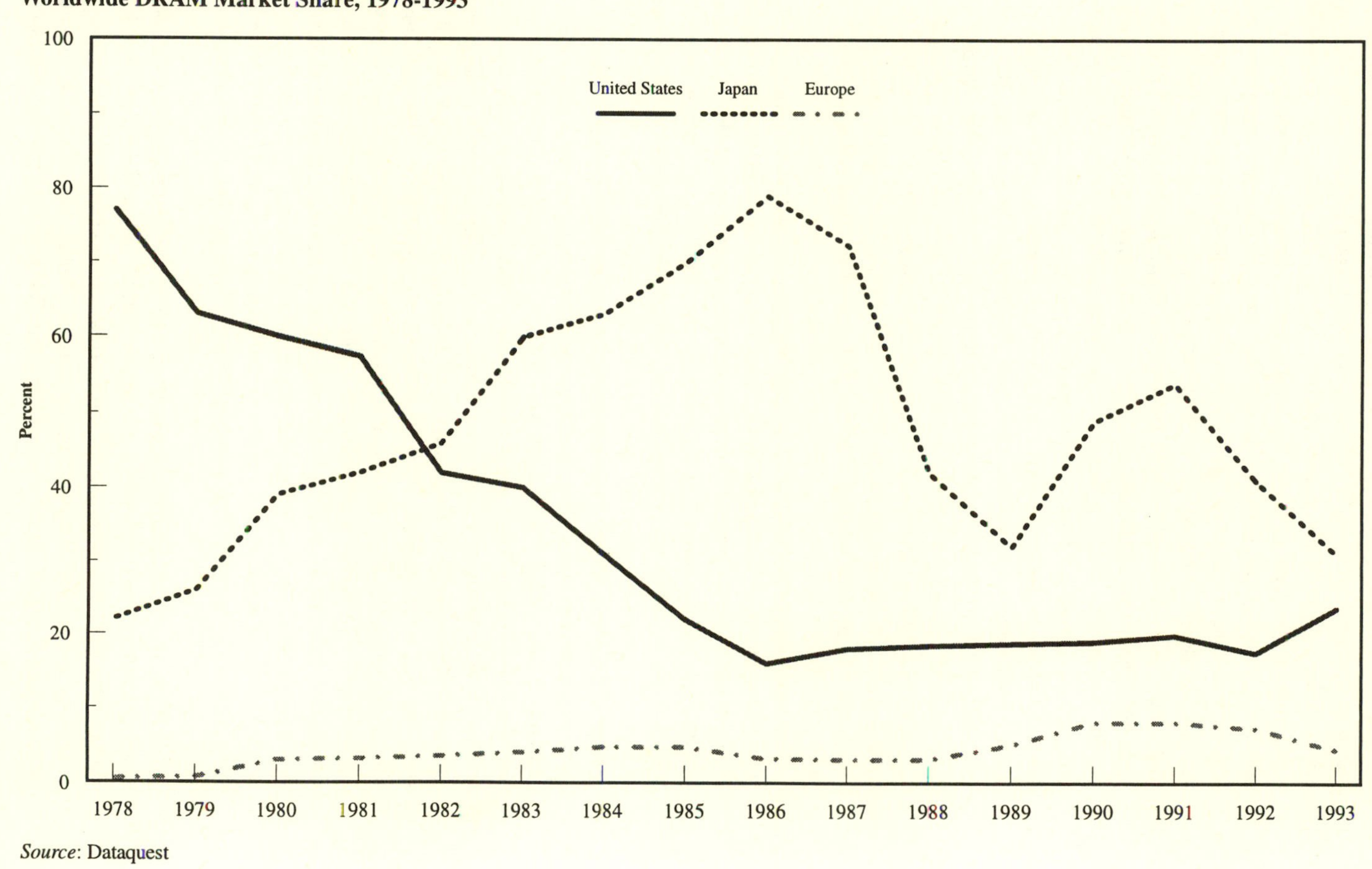

Source: Dataquest

duction-process technology that accelerated technological advances in a broad range of process and manufacturing areas that helped Japanese firms break into less standardized markets in the 1980's.

Had the Japanese government targeted microprocessors or ASICs (Application specific integrated circuits) they would have had less success because these are not commodity products. Competition for these components depends on proprietary design, and American firms had an advantage since their products had already been adopted as standards in important systems. By contrast, DRAMs give them knowledge in commodity manufacturing that dominates the world market for semiconductors. Through the 1980's, standard commodity devices accounted for 85% of the total market for semiconductors.[77] Japanese success in commodity memory markets permitted them to expand into other complicated commodity products such as EPROM. Although the United States maintained a lead in the more sophisticated custom design and device sectors, the loss of commodity technical leadership paved the way for the ascendancy of Japan as a world leader in the semiconductor industry.

Japanese Semiconductors Industrial Policy: 1980's and After

The face of Japanese industrial policy for semiconductors has changed in light of the new competitive environment. That environment is characterized by rough technological parity with the United States, a leading share of the world market, and the industrial maturity of those firms that MITI programs helped become competitive in the semiconductor field—enterprises that have become large, stable, and highly profitable players in the international arena. Table 5.1 indicates that four of the top five, and six of the top ten, semiconductor vendors worldwide are Japanese. Hitachi alone, for example, spends four times more on R&D than does MITI.[78] Although MITI still exerts some influence, the need to support these international giants has become less necessary.

The new emphasis is on the strategic guidance of future-generation technologies. Extending the technology frontier by establishing breakthrough technologies is the burden of technological parity with the United States, and the government is of the view that the private sector, even in Japan, lacks the long-term commitment necessary for success.[79] New themes in government promotion include an emphasis on long-term, high-risk technologies that offer potential breakthroughs rather than the catchup character of government policy in the 1970's. Government-sponsored R&D projects in the 1980's have had a long-term time horizon, the result of which has not yet reached commercial fruition.

MITI efforts in R&D have extended beyond channeling private-sector energies toward strategic commercial applications. MITI offers direct support in two forms. The government provides direct financial assistance by securing grants from the Ministry of Finance and channeling funds from some of its satellite organizations such as the Japan Development Bank, the JKTC, and NEDO (New

Table 5.1
Top Ten Semiconductor Vendors Worldwide
(millions of $)

Company	1978	Company	1989
Texas Ins	990	NEC*	4,964
Motorola	720	Toshiba*	4,889
NEC*	520	Hitachi*	3,930
Philips^	520	Motorola	3,322
National	500	Fujitsu*	2,941
Fairchild	500	Texas Ins	2,787
Hitachi*	460	Mitsubishi	2,629
Toshiba*	400	Intel	2,440
Intel	360	Matusushita	1,457
Siemens	270	Philips^	1,690

*Japanese Firm
^ European Firm

Source: Dataquest. Cited in *Electronic Buyers' News*, January 8, 1990.

Energy Development Corporation).[80] The government provides direct technological support by enlisting the efforts of national laboratories that are administered by MITI's Agency for Industrial Science and Technology (AIST). In addition, NTT continues to play an important role in technology development through the operation of a research laboratory devoted exclusively to semiconductor R&D.

During the 1980's the Japanese industry strengthened its position against European and American competitors. European firms were marginalized further from world competition and American firms lost market share to Japan in all major semiconductor subsectors. Figure 5.4 indicates that the United States global share in microcomponents is still greater than that of Japan, but it declined from 78% in 1980 to 59% in 1991. Japanese market share increased from 22 to 38% during the same period. Figure 5.5 reflects a similar market convergence for ASICs In Figure 5.6, the same convergence is reflected in another commodity memory, EPROM. All these trends are captured in Figure 5.7, depicting Japanese dominance of semiconductor world market share.

REPRISE

Japan's spectacular achievements in semiconductors have been facilitated by government intervention. However, R&D spending or protectionist measures are not any guarantee of technological and market leadership. European governments have invested greater sums in semiconductor and related R&D and still trail

Figure 5.4
Worldwide Microcomponent* Market Share, 1980-1993

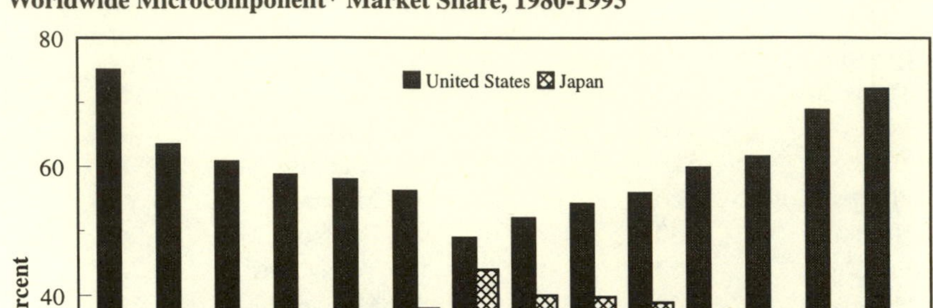

*Includes microprocessors, microcontrollers, and microperipherals

Figure 5.5
Worldwide ASICs Market Share, 1984-1993

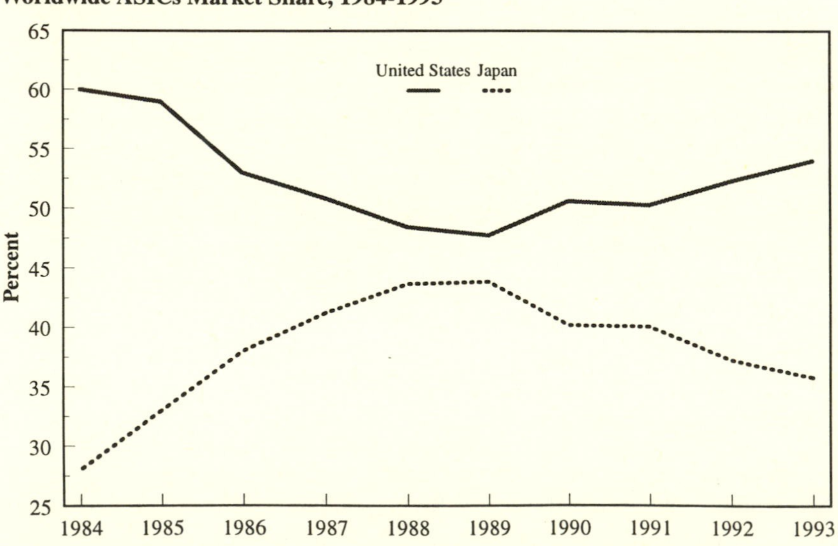

Source: Dataquest

Figure 5.6
Worldwide EPROM Market Share, 1978-1993

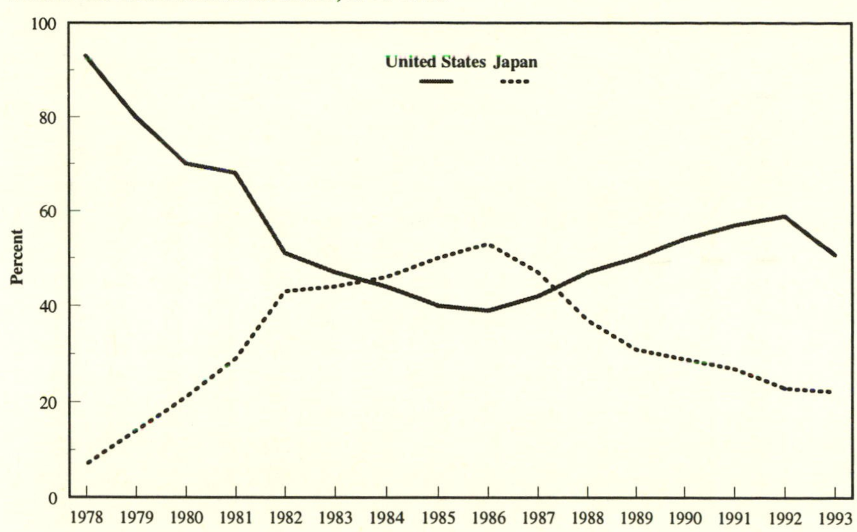

Source: Dataquest

Figure 5.7
Semiconductor World Market Shares, 1982-1993

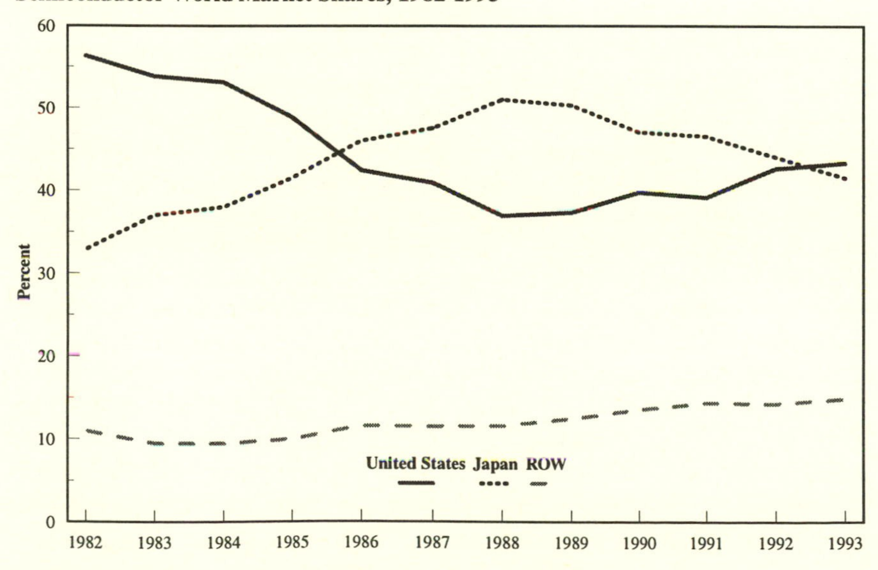

Source: Dataquest

badly in most areas. If the Japanese success is due to government intervention, how can one explain the persistent technology gap experienced by Europe after years of promotion? The answer must not only address the milieu in which the promotion was implemented but determine where and why policy has yielded the greatest rewards.

The economic and political milieu has been a source of strength for Japanese industrial policy. Political stability represented by the unbroken reign of the Liberal Democratic Party has brought uninterrupted progress in the industrial policy regime, permitting the government to implement long-term strategies of ten years or more.[81] The EC Commission commented in 1990 that "what distinguishes the Japanese implementation of industrial policy is its constancy and continuity compared with the frequent policy changes and reversals characteristic of Western governments."[82] By determining the strategic direction of the industry through collaboration with business and science leaders, MITI has been able to deploy resources to that sector directly, but more important, it has encouraged private agents to take part in a sector designated by the government to be a winner. Cooperative R&D projects, therefore, have an important signaling affect. It is arguably not the level of government funding that matters but the communication that a specific research activity is of future importance to the Japanese economy. Japanese semiconductor companies spend far more internally on private research and development in the same field than they contribute to a cooperative venture. In this way, MITI signaling continues to exert some influence on the R&D agenda.[83]

The business and industrial structure lends itself to government-industry cooperation, and Japanese industrial policy has utilized this relationship effectively in the implementation of industrial strategies. A large part of the Japanese economy is organized into *keiretsu*, large industrial groups sharing the integration of diffuse enterprises in a large, cooperative entity that competes against other similar organizations.[84] All the major Japanese semiconductor companies are tied to a *keiretsu*. These large, vertically integrated industrial groups buy and distribute among member firms and resist outside penetration of its structure. The Office of Technology Assessment stated that "keiretsu feelings are very strong and that members go against the group only when they feel their very survival depends on it."[85] They are joined together through cross shareholdings, interlocking directorates, and a web of personal connections. The heart of the grouping consists of a bank. These links reinforce a "buy Japan" bias and limit semiconductor import penetration. In addition, the *keiretsu* structure provides internal markets for semiconductors such that 45 to 80% of semiconductor output is consumed within the Group.[86] This concentrated structure lends itself to consensus building by the government and facilitates collaborative arrangements and efforts to draw resources into targeted sectors.

The evolution of the semiconductor industry in Japan and Europe has certainly been influenced by macro factors that also include the quality of the scientific and technological infrastructure, labor, the type of financial system, and the overall size and sophistication of market demand. Self-sufficient in semiconductors through the 1950's,[87] European industry was unable to meet growing challenges, reflected in its being surpassed in technological capability and market share by Japan in the 1970's. The failure of the European industry to keep pace and the success of Japan in catching up to America can be attributed to a combination of sector-specific factors. These include the organization of R&D, the production strategies of firms, the composition of demand, and the type and extent of government policy.

The European and Japanese semiconductor promotion strategies shared similar characteristics. Both relied on home-market protection through trade barriers and preferential procurement practices that guaranteed local producers a favorable market. Yet protection proved to be a failure in Europe while it appeared to have succeeded in Japan. A major reason why trade protection failed to create a world-class European semiconductor industry is the substantial foreign direct investment by American and later Japanese semiconductor houses. The Europeans even encouraged the development of a foreign-owned production base in Europe as a compromise solution to the competitive problems that high-priced semiconductor imports placed on European electronic systems producers. This enabled foreign producers to preempt the developing market for advanced semiconductor products that European houses were unable to supply. Foreign direct investment enabled American firms to effectively circumvent tariff barriers. American firms used technological assets to solidify production and marketing arrangements that would guarantee them a sizable market at the expense of European producers.

European trade policy has been generally less concerned with creating a European owned industry than in establishing a European-production base. Strategic concerns in the mid-1980's led to stricter rules of origin that were designed to expand the market for European producers; and encourage additional foreign direct investment to further strengthen the European production base. By intensifying the process to manufacture sensitive high technology inputs such as semiconductors within regional borders, European governments have eased national and economic security concerns, even if the manufacturers of those products are not predominantly European. By contrast, the Japanese objective was to create an indigenous semiconductor industry. Tight investment restrictions accompanied home-market protection to ensure that domestic demand would be reserved for Japanese companies. A beneficial and perhaps unintended result was the transfer of American technology. American companies did not have to transfer technology to European companies because of the more desirable option of productive investment.

A more important determinant of the success of home-market protection was the lack of competition in the European industry and the intensity of competition in the Japanese industry. European national champion policies discouraged competition and indirectly promoted large, inefficient firms unable to survive the standards of international competition if not for subsidy and preferential procurement. The Japanese government also ensured preferential procurement policies for local semiconductor houses, but did not pursue a national champion philosophy. An important factor in Japanese success in semiconductors was that industrial policies were implemented in a competitive environment. One senior MITI official observed that "Japanese are so locked into their own company (group) and are so competitive that they go beyond the bounds of normal economic behavior and engage in excessive competition with each other as much or more than with foreigners."[88] To avoid this excessive competition, MITI helped determine the level of entry into the semiconductor industry.[89] Nevertheless, intense domestic rivalries characterize the industry in Japan.[90] Competitive conditions have imposed discipline in the industry, fostered innovation, and drive companies to commercialize cooperative research results ahead of the competition. This encouraged the protected industry to develop efficiently despite limited international competition.

Japanese industrial policy was more coherent and consistent than the European policy, which differed in timing and scope. Both European and Japanese industrial policy for semiconductors has been linked to the promotion of their respective computer industries. European industrial promotion for semiconductors was top-down through the subsidization of large computer and electronic systems producers. Japanese industrial policy also granted subsidies for the industry, but emphasized a bottom-up approach, a long-term vision that linked the production of advanced semiconductor chips to international strength in the computer industry. One might have expected the increased competition from American and Japanese companies in the European market would encourage a more competitive European semiconductor industry. It did not for several reasons. First, national champion policies were anticompetitive. Second, unlike Japanese companies, European firms did not commit themselves to LSI technology in the early 1970's. Therefore, the telecommunications and computer industry had to shop elsewhere for LSI devices that were forming the core of new electronic systems. By contrast, when computer demand for advanced semiconductors increased during the late 1970's, the Japanese semiconductor industry was able to supply all but the most advanced technology to the domestic computer industry. The Japanese industry was able to build upon LSI technology in the government-sponsored VLSI program while Europe, just committing to LSI device production, lacked similar cumulative technology and expertise to mount a credible challenge. The combination of demand-pull and technology-push was an important ingredient in the success of the Japanese semiconductor industry.

Government policy has clearly had a strong influence on the course of semiconductor development in Europe and Japan, but the blend has been different and the outcomes mixed. In Japan, government coordination and investment of R&D activity, the continuous build up of an advanced technology capability through the accumulation of productive experience in LSI technology, demand factors, and a coherent and protective public policy were the basis for

Japan's ascendance in the industry. In Europe, the importance of an indigenous semiconductor capability became more obvious to European producers and governments by the late 1970's and early 1980's. A growing European foreign semiconductor dependency became less tenable with the precipitous decline in the fortunes of local producers and the ascendancy of Japanese producers on the world market. The fear of further marginalization in an important industry[91] and heightened fears over access for downstream markets encouraged a new orientation in semiconductor promotion. Past policy failures have not deterred European governments from using public resources to strengthen their semiconductor industry in the 1980's and 1990's, the results of which are yet to be completely assessed.

NOTES

1. U.S. Department of Commerce, *Report on the U.S. Semiconductor Industry* (Washington, D.C.: Government Printing Office, 1979), 11.

2. Many American semiconductor producers shifted the lower value labor-intensive assembly of semiconductor devices to offshore plants in low-wage countries. This cost-cutting maneuver, estimated to be around 50%, began in 1964 and by 1979 there were over 89,000 people employed in such assembly operations compared to 85,000 in the United States. Import law TSUS 806 and 807 stipulated that the firm had to pay duty only on the value added offshore. From 1969 to 1979, 80% of semiconductor imports to the United States represented the reimportation of output from domestically headquartered companies. Since 1979, this figure has decreased, reflecting the expanding strength of Japanese semiconductor exports. U.S. International Trade Commission, *Foreign Industrial Targeting and Its Effects on U.S. Industries, Phase I: Japan*, USITC Publication no. 1437 (Washington, D.C.: Government Printing Office, 1983), 141. For effects on semiconductor employment, see Carol A. Parsons, "The Changing Shape of Domestic Employment in High-Tech Industry: The Case of International Trade in Semiconductors," in *The Dynamics of Trade and Employment*, Laura D'Andrea Tyson, ed. (Cambridge: Ballinger Publishing Company, 1988), 237-242.

3. The Basic Technology Research Act provided fiscal and tax incentives for R&D, funding for HDTV, flat panel displays, etc., and the Japan Key Technology Center was established to fund high-risk high technology R&D whose commercial fruition would be anywhere from 5 to 15 years in the future.

4. Semiconductor Industry Association, *Fourth Annual Report* (Cupertino: Semiconductor Industry Association, 1990).

5. U.S. National Advisory Committee on Semiconductors, *A Strategic Industry at Risk* (Washington, D.C.: Government Printing Office, 1989), 11.

6. WSTS (World Semiconductor Trade Statistics) cited in Thomas R. Howell, Brent L. Bartlett, and Warren Davis, *Creating Advantage: Semiconductors and Government Industrial Policy in the 1990's* (San Francisco: Semiconductor Industry Association, 1992), 9.

7. European Communities Commission, *The European Electronics and Information Technology Industry: State of Play, Issues at Stake, and Proposals for Action* (Luxembourg: Commission of the European Communities, March 1991), 28-36.

8. Europe viewed the STA with some concern because Japan and the United States agreed bilaterally on a set of conditions that could affect the availability and price of a product of strategic importance to Europe. It illustrated that power in the industry stemmed from the production side, and that Europe was at a strategic disadvantage by not having a world-class industry. At the same time, Europe was concerned that cheaper and more reliable Japanese components would decimate segments of their industry as it had done in America. See Kenneth Flamm, "Semiconductors," in *Europe 1992: An American Perspective*, Gary Clyde Hufbauer, ed. (Washington, D.C.: The Brookings Institution, 1990), 257-259. See also "Economics of Managed Trade," *The Economist*, 22 September 1990, 19.

9. Variations in end-user demand leads to an uneven semiconductor demand. For example, in 1984 demand skyrocketed, reflecting the replenishment of computer inventories, resulting in a 55% increase in semiconductor imports. Slack computer demand the following year drove semiconductor imports down 31%. Variations in demand reflect those end-user markets being served. See Laura D'Andrea Tyson, "The Changing Shape of Domestic Employment in High-Tech Industry: The Case of International Trade in Semiconductors," in *The Dynamics of Trade and Employment*, Laura D'Andrea Tyson, ed. (Cambridge: Ballinger Publishing Company, 1988), 241.

10. Korea announced a semiconductor industry promotion plan in 1982. Between 1990 and 1996, government planned to spend $2 billion and increase capital investment by $3.4 billion with tax benefits and regulations. Procurement and tariffs further promoted Korean producers. Thomas R. Howell, Brent L. Bartlett, and Warren Davis, *Creating Advantage: Semiconductors and Government Industrial Policy in the 1990's* (San Francisco: Semiconductor Industry Association, 1992), 348. Taiwan mapped out a similar high technology strategy in the mid 1980's. See Walter Arnold, "Science and Technology Development in Taiwan and South Korea," *Asian Survey* (April 1988).

Korea has made major inroads into the DRAM market. Samsung has emerged as one of the top producers of DRAMs and has focused on expanding market share by underselling many rival Japanese producers. Part of that effort has been focused on the American market. In October 1992, the commerce department imposed anti-dumping duties on Samsung and Goldstar DRAM chips—87.4% and 52%, respectively. Many analysts believe that if a March 1993 ruling confirms those penalties, it will effectively wipe Samsung and Goldstar out of the American market and deal a debilitating blow to their aspirations in the industry. The Korean semiconductor industry has 3.8% of world market share. Andrew Pollack, "A Chip Powerhouse is Challenged," *New York Times*, 17 December 1992, D1.

11. Prior to 1978, aid for the semiconductor industry was channeled mostly through the domestic computer industry and R&D subsidies used in the electronic components industry. In 1978 the government created a four-year plan for the semiconductor industry to develop a manufacturing capability in leading integrated circuits, but was unable to close the gap with the American and Japanese industry owing to long delays in the implementation of the plan, support unequal to the R&D, and required capital investments. See Franco Malerba, *The Semiconductor Business: The Economics of Rapid Growth and Decline* (London: Frances Pinter, 1985), 194.

12. Le Monde (22 September 1982) commented: "losses are expected but . . . memories must be manufactured in France. . . . Even if it means losing our shirts on it, it is the unanimous affirmation of all the specialists. . . . Conditions are favorable provided there be no lack of public money." Cited in Thomas R. Howell, Brent L.

Bartlett, and Warren Davis, *Creating Advantage: Semiconductors and Government Industrial Policy in the 1990's* (San Francisco: Semiconductor Industry Association, 1992), 326.

13. The plan suffered technological exclusion from potentially rewarding international alliances. Investment levels were less than anticipated by the government; the government underestimated the vigor of American and Japanese completion, and the price war in chips left the nationalized firms with huge operating losses such that in 1984 the government bailed them out to the tune of over $1 billion. Ibid.

14. In 1991 the government was prepared to invest another $800 million in equity in both SGS-Thomson and Groupe Bull. Johnathon Levine, "European High Technology Tries to do Something Drastic: Grow Up," *Business Week*, 25 March 1991, 48.

15. Thomson became SGS-Thomson in 1987 when the French Thomson merged with Italian SGS to form the world's 12th largest producer of semiconductors. SGS was a national semiconductor champion of Italy.

16. In 1978 the Labor government instituted the Microelectronics Support Program, giving $90 million in investments and R&D subsidies to semiconductor firms. The Microprocessor Applications Program added $130 million in grants for the diffusion of microelectronic technology.

17. The Microelectronics Support Program II was adopted (1984-1990) with a budget of $200 million. In 1988 the LINK program was conceived to promote the commercial applications of government-funded R&D projects between scientists and firms.

18. "It has to be faced that most of the products requiring capital investment—that is most of the hardware projects—did not produce anything which has yet reached the market; nor are they ever likely to do so," *New Technology Week*, 4 March 1991. Cited in Thomas R. Howell, Brent L. Bartlett, and Warren Davis, *Creating Advantage: Semiconductors and Government Industrial Policy in the 1990's* (San Francisco: Semiconductor Industry Association, 1992), 346.

19. German aid to depressed industries such as coal mining, textiles, and clothing is the lowest in the EC. Bernard Uddis, *The Challenge to European Industrial Policy: Impacts of Redirected Military Spending* (London: Westview Press, 1987), 62.

20. According to interviews conducted by Malerba, the government effectively financed 20% of all R&D expenditures conducted by Siemens in integrated circuits through the 1970's. Franco Malerba, *The Semiconductor Business: The Economics of Rapid Growth and Decline* (London: Frances Pinter, 1985), 189.

21. Includes telecommunications, semiconductors and related electronic components, electronic consumer goods, computers, and industrial automation.

22. Thomas R. Howell, Brent L. Bartlett, and Warren Davis, *Creating Advantage: Semiconductors and Government Industrial Policy in the 1990's* (San Francisco: Semiconductor Industry Association, 1992), 337.

23. Major FHG institutes for microelectronic research include FHG for Silicon Technology, FHG Institute for Microstructure Techniques, FHG Institute for Integrated Circuits, FHG Institute for Microelectronis Circuits and Systems, and The Society for Silicon Applications and CAD/CAT. The government has provided assistance in the area of CAD integrated circuits ($235 million from 1986-90), the Gallium Arsenide Project ($60 million since 1986), Supercomputer Consortium ($115 million 1986-90), and HDTV ($45 million 1990-93). The government typically funds 50% of program expenses. Ibid., 337-340.

24. Ingolf Ruge of the FHG Solid State Technology offered: "The goal of the

Japanese . . . and they do have a goal for everything . . . is a world monopoly on chips. . . . With that the Japanese have access to all modern technologies. They have control over what will be the most important economic asset in the year 2000." Der Spiegel (24 April 1989), Cited in Ibid., 334.

25. The source of Europe's shortcomings in the general high technology field stem from the lack of a truly integrated market; uncoordinated standards and procedures; redundancy in R&D; the lack of harmonized markets downstream and the lack of coordination in technological development upstream; a lack of venture capital; and the relatively small amount of military funds spent on R&D that have helped major American industrial firms accumulate strength and expertise they otherwise would not have enjoyed. See Hubert Curien, "The Revival of Europe," in *A High Technology Gap? Europe, America, and Japan*, Frank Press, Hubert Curien, and Keichi Oshima, eds. (New York: Council on Foreign Relations, 1987), 44-47. See also P. R. Beije and others, *A Competitive Future for Europe? Towards A New European Industrial Policy* (London: Croom Helm, 1987).

26. Organization of Economic Cooperation and Development, *Economic Outlook* (Paris: OECD, 1975), 16.

27. WSTS data in Kenneth Flamm, "Semiconductors," in *Europe 1992: An American Perspective*, Gary Clyde Hufbauer, ed. (Washington, D.C.: The Brookings Institution, 1990), 236.

28. The LSI period lasted from 1971 to the early 1980's, at which time the VLSI (very large scale integration) period began. The LSI and VLSI technologies were not radical departures, but represented an extension of previous integrated circuit technology, one that differed in the degree of integration and miniaturization of semiconductor devices. By contrast, the transistor technology was a radical departure from previous vacumn technology, and the integrated circuit established a radically different technological regime than the one with preceding transistors.

29. European firms concentrated on producing specialized proprietary chips to serve niche markets in the community and did not develop advanced manufacturing skills critical to success in commodity memory products. Memories became the largest segment of the global semiconductor industry in the 1980's, and European firms were not in a position to compete with American and Japanese producers. Ibid., 187.

30. While the large electronic producers (Philips, Siemens, Thomson-CSF, and AEG-Telefunken) purchased state-of-the-art chips from American companies it was recognized that a capability in advanced-integrated circuit technology was important for long-term competitiveness in electronic end markets. For example, during peak demand American producers preferred to supply American customers before European customers. The large national producers endeavored to create an in-house capability and sought to catchup through the creation of large-scale integrated capabilities subsidized from other areas of the firm; acquisition of American merchant producers at the frontier of technology; and with government support. Giovanni Dosi, "Semiconductors: Europe's Precarious Survival in High Technology," in *Europe's Industries: Public and Private Strategies for Change*, Geoffrey Shepard, ed. (Ithaca: Cornell University Press, 1983), 57.

31. Michael Borrus, *Competing for Control: America's Stake in Microelectronics* (Cambridge: Ballinger, 1988), 78.

32. American merchants transferred the technology to Europe as exports or for production in the local subsidiaries only after the American merchants had gained cost

and quality advantages through learning and scale economies. This gave them advantages in quality and cost that European producers could not overcome. Quick, Finan, and associates, *International Transfer of Semiconductor Technology through U.S. Based Firms*, NBER Working Paper 118 (New York: National Bureau of Economic Research, 1975), 49.

33. This happened after 1987 when MITI production guidelines restricted DRAM production enough to boost prices substantially, leading in the late 1980's to what was referred to by industry insiders as "bubble money." Kenneth Flamm, "Semiconductors," in *Europe 1992: An American Perspective*, Gary Clyde Hufbauer, ed. (Washington, D.C.: The Brookings Institution, 1990), 257.

34. Thomas R. Howell, Brent L. Bartlett, and Warren Davis, *Creating Advantage: Semiconductors and Government Industrial Policy in the 1990's* (San Francisco: Semiconductor Industry Association, 1992), 117-131.

35. One American firm lost its supply of DRAM's when its negotiations with the Japanese producer on an unrelated matter reached an impasse. Thomas R. Howell, Brent L. Bartlett, and Warren Davis, *Creating Advantage: Semiconductors and Government Industrial Policy in the 1990's* (San Francisco: Semiconductor Industry Association, 1992), 127.

36. U.S. General Accounting Office, *U.S. Business Access to Certain Foreign State-of-the-Art Technology* (Washington, D.C.: Government Printing Office, 1991), 43-45.

37. *L'Usine Nouvelle*, 2 January 1992. The same engineer stated that it is very difficult to challenge the Japanese in HDTV while being dependent on key components of HDTV development. Cited in Thomas R. Howell, Brent L. Bartlett, and Warren Davis, *Creating Advantage: Semiconductors and Government Industrial Policy in the 1990's* (San Francisco: Semiconductor Industry Association, 1992), 130.

38. U.S. General Accounting Office, *U.S. Business Access to Certain Foreign State-of-the-Art Technology* (Washington, D.C.: Government Printing Office, 1991), 29-30. It was indicated that Japanese access to the most advanced displays is assured and in a timely fashion.

39. Strategic alliances in aerospace (Concorde, Airbus) a prominent exception.

40. EEC Regulation 288/89, in European Communities Commission, *Official Journal of the European Communities*, no. L33 (Luxembourg: EC Commission, February 4, 1989), 23.

41. EC Commission directives governing the award of government contracts stipulate that contractors may ignore bids in which half of the value of the contract is of EC origin. Since public-sector purchases of telecommunications equipment can account for 90% of sales for telecommunications companies by American firms in Europe, and 33% for computer firms, the rules of origin is a source of concern. Kenneth Flamm, "Semiconductors," in *Europe 1992: An American Perspective*, Gary Hufbauer, ed. (Washington, D.C.: The Brookings Institution, 1990), 270.

42. Roger Cohen, "Europe's State-Industry Ties: Success and Utter Failures," *New York Times*, 8 November 1992, 1.

43. Margaret Sharp, "The Single European Market and European Technology Policies," in *Technology and the Future of Europe: Global Competition and the Environment in the 1990's*, Christopher Freeman and Margaret Sharp, eds. (London: Francis Pinter, 1991), 62.

44. DTI, *Guide to European Community Industrial R&D Programs*, (July 1990), cited

in Margaret Sharp, "The Single European Market and European Technology Policies," in *Technology and the Future of Europe: Global Competition and the Environment in the 1990's*, Christopher Freeman and Margaret Sharp, eds. (London: Francis Pinter, 1991), 66.

45. Ibid., 73.

46. RACE: Conceived with the goal of establishing the standards and technological base for a European Integrated Broadband Communications Network, with the objective of creating a unified telecommunications network in the European Community. Specific technological areas include high-performing integrated circuits, broad-band switching, passive optical components, image coding, and technology related to large flat-screen displays. During its five-year mission (1988 to 1992), RACE was expected to assist in the development of HDTV with over 120 participating companies in this program. BRITE: Conceived in 1982 by the commission to stimulate technological development in advanced fields. It is the functional equivalent of ESPRIT in the technical fields, such as new materials and production techniques not covered by ESPRIT. Its primary role is one of catalyst, a clearinghouse of ideas and interests expressed by industry as to appropriate avenues of technological development to pursue. COMMETT: Conceived in 1985, with the purpose of integrating European research structures by facilitating the exchange of scientists and students between private enterprises and universities with scholarships. Stimulation: Addresses the needs of basic research. Goal is to stimulate cooperation and coordination among the European scientific community.

47. In addition, insistence by the EC that even small and medium-sized companies be permitted to partake in the program and the results of the research bothered larger firms and the larger member governments that viewed such a setup would compromise their ability to catch up to Japan and the United States. ESPRIT Review board, *The Review of ESPRIT 1984* to 1986 (Luxembourg: Commission of the European Communities, May 1989), xii.

48. Governments have at times resisted or disagreed on levels of funding. One such disagreement resulted in the partial marriage of the EUREKA program in semiconductors (JESSI) with the efforts of ESPRIT, with the EC program providing some funds and undertaking some of the R&D previously consigned to JESSI.

49. 1 ECU =@. $1.2.

50. Efforts by European companies such as Philips and ASM International to participate in SEMATECH were refused. These and other European companies tied IBM Europe participation in JESSI contingent upon their entry into SEMATECH, but IBM Europe defused the conflict by offering access to its own X-ray lithography program and by agreeing to join forces with Siemens on 64 megabit DRAMS. See Kenneth Flamm, "Semiconductors," in *Europe 1992: An American Perspective*, Gary Clyde Hufbauer, ed. (Washington, D.C.: The Brookings Institution, 1990), 283.

51. Quote in Hubert Curien, "The Revival of Europe," in *A High Technology Gap? Europe, America, and Japan*, Frank Press, Hubert Curien, and Keichi Oshima, eds. (New York: Council on Foreign Relations, 1987), 61.

52. JESSI Planning Committee, *Jessi Program—Results of the Planning Phase* (Luxembourg: Commission of the European Communities, February 1990), 39.

53. ESPRIT Review Board, *The Review of ESPRIT 1984-1988* (Luxembourg: Commission of the European Communities, May 1989), 5.

54. Ibid., 7.

55. Barbara Berkman, "Start-Up No More, Jessi Gets Down to Business," *Electronic*

Business, 18 May 1992, 68-72.

56. Ibid., 70.

57. *Financial Times*, 7 February 1985, cited in Thomas R. Howell, Brent L. Bartlett, and Warren Davis, *Creating Advantage: Semiconductors and Government Industrial Policy in the 1990's* (San Francisco: The Semiconductor Industry Association, 1992), 324.

58. Roger Cohen, "Europe's State-Industry Ties: Success and Utter Failures," *New York Times*, 8 November 1992, A1.

59. European firms enjoy only 10% of the world semiconductor market, 15% of the world market for computer peripherals, and 20% of the world consumer electronics market. With regard to computers ($45 billion computer market), European computer production satisfies only 66% of European demand (U.S. subsidiaries serve 60% of European market). See European Communities Commission, *The European Electronics and Information Technology Industry: State of Play, Issues at Stake, and Proposals for Action* (Luxembourg: Commission of the European Communities, March 1991), 17-26.

60. John R. Blau, "Europe Stumbling in Semiconductor Race," *Research and Technology Management* (March/April 1992), 4.

61. Mike Hobday, "The European Semiconductor Industry: Resurgence and Rationalization," in *Technology and the Future of Europe: Global Competition and the Environment in the 1990's*, Christopher Freeman and Margaret Sharp, eds. (London: Francis Pinter, 1991), 86.

62. In 1990 Siemens and IBM created a partnership to create 64M DRAMs by 1994. This could make Siemens a world leader in DRAM technology.

63. The Japan Electronic Computer Company (JECC) was established in 1961 by a joint venture among the leading Japanese computer producers. From 1961 to 1991, the Office of Technology Assessment indicated that the government had channeled $6 billion in low-interest loans to the JECC. See U.S. Office of Technology Assessment, *Competing Economies: America, Europe, and the Pacific Rim* (Washington, D.C.: Government Printing Office, 1991), 261.

64. In 1960 IBM was allowed to set up a 100% owned manufacturing subsidiary in Japan in return for a commitment to enter into licensing agreements with 13 Japanese companies. This was the only wholly owned manufacturing subsidiary permitted by MITI during these years. In semiconductors, MITI approved a joint venture between Texas Instruments (TI) and Sony in 1968, and in 1972 to produce bipolar circuits. TI bought out Sony's share and remained the only wholly owned manufacturing subsidiary in Japan until 1980.

65. U.S. Federal Trade Commission, *Staff Report on the Semiconductor Industry* (Washington, D.C.: Government Printing Office, 1989), 86.

66. Averaged 10% of sales through the 1970's. Ibid., 148; U.S. National Research Council, *U.S.-Japan Strategic Alliances in the Semiconductor Industry: Technology Transfer, Competition, and Public Policy* (Washington, D.C.: National Academy Press, 1992), 13.

67. Daniel I. Okimoto, *Between MITI and the Market: Japanese Industrial Policy for High Technology* (Stanford: Stanford University Press, 1989), 27.

68. Investment restrictions were lifted in 1978. Quick, Finan, and Associates (1985), cited in Laura D'Andrea Tyson, *Who's Bashing Whom: Trade Conflict in High Technology Industries* (Washington, D.C.: Institute for International Economics, 1992), 93.

69. Japanese R&D programs were insignificant at this time compared to those

undertaken in the United States. For example, in 1970, private company funding of semiconductor and computer R&D of the three leading electronic houses, NEC, Hitachi, and Fujitsu, was less than the R&D budget of Texas Instruments. Ira Magaziner and Thomas Trout, *Japanese Industrial Policy* (Berkeley: Institute of International Studies, 1981), 103.

70. Daniel I. Okimoto, *Between MITI and the Market: Japanese Industrial Policy for High Technology* (Stanford: Stanford University Press, 1989), 162.

71. For example, NEC and Hitachi specialized in INMOS and CMOS LSI devices, Fujitsu in NOMOS memory and bipolar logic, Toshiba in CMOS LSI devices, and Matsushita in linear integrated circuits and selected memory devices. See Michael Borrus, James Millstein, and John Zysman, *U.S.-Japanese Competition in the Semiconductor Industry: A Study in International Trade and Technological Development* (Berkeley: Institute of International Studies, 1982), 66.

72. Japanese firms had already cornered 85% of the American calculator market by 1971. U.S. Office of Technology Assessment, *Competing Economies: America, Europe, and the Pacific Rim* (Washington, D.C.: Government Printing Office, 1991), 246.

73. U.S. International Trade Commission, *Foreign Industrial Targeting and Its Effects on U.S. Industries, Phase I: Japan*, USITC Publication no. 1437 (Washington, D.C.: Government Printing Office, 1983), 98.

74. U.S. Department of Commerce, *Report on the U.S. Semiconductor Industry* (Washington D.C: Government Printing Office, 1979), 82.

75. The project was nonprofit, designed to enable members to compete more quickly and more effectively with competition. Main participants were Hitachi, Fujitsu, NEC, Oki, and Toshiba. See U.S. General Accounting Office, *Industrial Policy: Case Studies in the Japanese Experience* (Washington, D.C.: Government Printing Office, 1982), 9.

76. Thomas R. Howell, Brent L. Bartlett, and Warren Davis, *Creating Advantage: Semiconductors and Government Industrial Policy in the 1990's* (San Francisco: Semiconductor Industry Association, 1992), 12.

77. Dataquest, cited in Michael Borrus, *Competing for Control: America's Stake in Microelectronics* (Cambridge: Ballinger, 1988), 158.

78. The MITI R&D budget through the 1980's has averaged $750 million. This was dispersed throughout all of industry, 18% of which went to high technology projects, 11% to microelectronic related R&D. Hitachi spends close to 4 billion on R&D. Neil Gross, "Inside Hitachi," *Business Week*, 28 September 1992, 93.

79. Japan Ministry of International Trade and Industry, *Vision of MITI Policies in the 1980's* (Tokyo: MITI, 1980), 136. See also Japan Ministry of International Trade and Industry, *Vision of MITI Policies in the 1990's* (Tokyo: MITI, 1990).

80. The Japan Key Technology Center (JKTC), Science and Technology Agency, and the Ministry of Education, Science, and Culture are other government agencies enhancing the competitive position of Japanese microelectronics. The JKTC is basically a government investment bank, conceived in 1985 with the purpose to provide capital and soft loans to high technology R&D. It is funded by private-sector sources and government and pool of funds are estimated in the billions of dollars. See Thomas R. Howell, Brent L. Bartlett, and Warren Davis, *Creating Advantage: Semiconductors and Government Industrial Policy in the 1990's* (San Francisco: Semiconductor Industry Association, 1992), 275. Under the aegis of the Science Agency is the Japan Research and Development Corporation (JRDC), Superconducting Material Research Multicore Project, Exploratory Research for Advanced Technology Organization (ERATO), Perfect

Crystal Project (1981 to 1986) and the TeraHertz Project (1987 to 1992), the Nano-Mechanism Project (1985 to 1990), and the Quantum Magneto Flux Logic Project (1989-93). Ibid., 278-83.

81. That has been challenged by the formation of a new coalition government in Japan, the first opposition government in Japan for 37 years. Although single-party rule promoted stability, in the end the LDP has failed to maintain power, in part because of a popular rejection of corrupt political-business practices that have occurred over the years of Japanese high technology industrialization.

82. European Communities Commission (DG IV), *Industrial Consequences of Targeting*, Working paper (Luxembourg: Commission of the European Communities, January, 1990), 14. Compare to Western governments where frequent change in party tenure occasion a reversal of previous policies and an inconsistent impact on economic policies and industrial structure.

83. See about soft policies of MITI in Ken-ichi Imai, "Japan's Industrial Policy for High Technology Industry," in *Japan's High Technology Industries: Lessons and Limitations of Industrial Policy*, Hugh Patrick, ed. (Seattle: University of Washington Press, 1986).

84. *Zaibatsu* refers to the industrial grouping of business around a financial house that in Japan resulted in the bifurcation of a modern led *zaibatsu* sector and a poorly organized and exploited medium and small business sector. Believing that these industrial groupings were responsible for Japan's military aggression, the Americans attempted to break up the zaibatsu networks after World War II. The onset of the Cold War compelled America to reassess this policy, and instead the United States encouraged the reformation of the zaibatsu along commercial lines in which managers worked closely with state officials to rebuild the economy. The new term assigned to this new but essentially unchanged industrial organization was *keiretsu.* There are three types of *keiretsu*, financial, distribution, and production *keiretsu.* The most common is the financial *keiretsu* in which the group revolves around a major commercial bank. The largest financial institutions in the world are *keiretsu* banks. These groupings also promote in-house procurement. This is part of the reason why the 20% market share for American semiconductors agreed to in the Semiconductor Trade Agreement of 1987 has been difficult to realize. See Masaru Yoshitomi, "Keiretsu: An Insider's Guide to Japan's Conglomerates," *International Economic Insights* (September/October 1990), 10-14. See also Chalmers Johnson, "Keiretsu: An Outsider's View," *International Economic Insights* (September/October 1990), 16-17.

85. U.S. Office of Technology Assessment, *Competing Economies: America, Europe, and the Pacific Rim* (Washington, D.C.: Government Printing Office, 1991), 261.

86. This industrial relationship also softens the impact that periodic falls in demand have on semiconductor operations and investment. See Michael Borrus, *Competing for Control: America's Stake in Microelectronics* (Cambridge: Ballinger, 1988).

87. Although European firms performed less well at the innovative level, they performed well at the productive and commercial levels. This was during the transistor period and before development of the integrated circuit consigned the American industry to a decidedly favorable position. See Peter Morris, *A History of the World Semiconductor Industry* (London: Frances Pereginus on behalf of the Institution of Electrical Engineers, 1990), 18.

88. Quoted in Hugh Patrick, "Japanese High Technology Industrial Policy in Comparative Context," in *Japan's High Technology Industries: Lessons and Limitation*

of Industrial Policy, Hugh Patrick, ed. (Seattle: University of Washington Press, 1986), 11.

89. MITI limited the number of firms permitted to license state-of-the-art integrated circuit technology from Texas Instruments in 1968. Through cooperative R&D projects the government has promoted large companies over small. Peter Morris, *A History of the World Semiconductor Industry* (London: Frances Pereginus on behalf of the Institution of Electrical Engineers, 1990), 103. See also *Japan Economic Journal*, 19 November 1968.

90. By one estimate there are 34 Japanese rivals in the semiconductor industry. Michael Porter, *The Competitive Advantage of Nations* (New York: The Free Press, 1990), 412.

91. Alain Gomez, chairman of Thomson, recently argued that the European industry must slap on "super-high and prohibitively strong" tariffs to ward off penetration of Japanese electronics to save the European electronics industry. He concedes that the consumer will be worse off, but offers "what does it mean to be a happy consumer if you are permanently out of work in a declining nation?" "The Case Against Free Trade," *Fortune*, 20 April 1992, 159.

6

SEMICONDUCTOR COMPETITIVENESS: THE UNITED STATES AND JAPAN

The preceding chapter detailed how semiconductors were promoted in Europe and Japan and the varied efficacy of those policies. This chapter will examine in more detail those forces shaping competitive advantage in trade and technology. Success in the industry is driven by technology, but it also depends on a breadth of serviceable markets, on a supporting semiconductor infrastructure, on healthy downstream industries, and on a sizable and steady supply of capital to cover escalating development costs. Semiconductor promotion as implemented in industrial structures different from the American free-market model is important to the dynamics of competition in the semiconductor industry. Government targeting may be an important element in competition, but it is unlikely that government can control all the factors shaping competitive advantage. Although the distortion of trade by government has had a profound influence on international competition, not all trade conflict stems from government targeting. This chapter examines the competitive position of the American industry in light of its principal rival, Japan. An overriding objective is to assess the extent to which America's decline in the industry stems from Japanese industrial policies or from other factors that have a large effect on semiconductor competition.

The semiconductor industry is one of the most capital—and technology—intensive industries in the world. The design and marketing of a single chip can run in excess of $50 million, and a state-of-the-art manufacturing facility may run as high as $800 million. Short product life cycles, typically two to three years, provide only a narrow window of opportunity to recoup the capital investment in product generation before the next product generation goes to market, and because per unit production costs decrease as volume of output increases, major competitive advantages can be gained by serving the widest range of markets. In addition, beating the competition to market provides a jump-start down the learning curve, which further enhances first-mover advantage.

COMMERCIAL PROMOTION IN THE UNITED STATES

Beyond the initial startup of integrated circuits in the early 1960's, government procurement and public policy have not played an important role in the evolution of the American industry. However, with the emergence of Japan as a major semiconductor producer there has been growing movement toward organizing a more coherent approach to competition. Since the early 1980's several organizations have come into existence and progress has been made toward defining a national technological agenda, and while the federal government has become a more willing participant in this effort, the primary initiative and impetus came from private industry.

The first two private cooperative ventures created to help the American industry compete more effectively were U.S. Memories and the Microelectronics and Computer Technology Corporation (MCC). Founded in 1983, U.S. Memories was a $500 million research consortium of leading industry players conceived to regain the competitive edge in DRAM production. However, despite substantial funding and industry support, the consortium fell apart in the mid-1980's owing largely to conditions of oversupply as plummeting revenues in the DRAM market compromised the ability of the consortium to regain lost market share. Formed in 1982, the MCC represented an American response to the Japanese VLSI program and an effort to avoid wasteful duplication of company research and development efforts in the industry. The for-profit consortium is made up of 25 firms overseeing and conducting research on integrated circuit technology, computer architecture, and software. The government has no role in this consortium. MCC has a budget under $100 million and approximately 450 employees, but only a few of the member companies have found it profitable to implement the limited technological results of MCC. Viewed in the context of a response to the Japanese VLSI program, it has been ineffective and insignificant. The largest industry players, IBM and AT&T, are not even consortium members. Critics consider the objectives ill-defined, and its agenda of research unclear and unfocused.[1] The MCC has transferred rather esoteric technologies to its member companies, but most observers and consortium participants concede that it has produced far too lean a return on the $372 million funneled into the MCC as of 1991. The significant effect of MCC and U.S. Memories is that they precipitated the enactment of the 1984 National Cooperative Research Act, a measure that exempted companies engaging in collaborative research from monopoly regulations and paved the way for R&D alliances and consortium-type technology development.[2]

In another private initiative members of the Semiconductor Industry Association (industry trade association) sponsored the creation of the Semiconductor Research Cooperative (SRC) in 1983. The cooperative consists of over 30 members (semiconductor device firms) and in most ways acts as a broker between member firms and universities to which it allocates funds to carry out its research projects. It is less involved with the extension of the technology frontier than with disseminating information and in providing a broad perspective on the best long-term research strategy for the semiconductor industry. By 1992 the SRC was funding $30 million in basic silicon-based research. The organization had relied exclusively on industry contributions for its funding requirements, but since 1986 three federal agencies have begun participating in the SRC and government aid over that five-year period has reached $9 million.[3]

Another component of the technological infrastructure of the semiconductor industry is the Defense Advanced Research Projects Agency, the primary federal organization entrusted with developing the necessary technology for defense purposes. The agency maintains a Microelectronics Technology Office (MTO) in which it conducts a variety of R&D projects. Despite some funding for dual-use technologies however, the overriding theme is military technology and the promotion of that technology for defense-related purposes.[4] The agency financed research in very high speed integrated circuits (VHSIC) and has sponsored the Monolithic Microwave and Millimeter Wave Integrated Circuit (MIMIC) project. Contrary to most European and Japanese initiatives, these programs have a primary military objective, the VHSIC to create integrated circuits that can sustain a more rapid performance rate in hazardous environments and the MIMIC to develop a gallium arsenide integrated circuit for VHSIC chips. By coincidence, one of the main goals of the VHSIC program, high-speed miniaturization, is also a goal shared by semiconductor producers in commercial applications, and some results of the VHSIC project have been used commercially.[5] The MIMIC technology, however, offers little to no advantages in commercial competitiveness of member firms.

The single most significant public or private effort to arrest the relative decline of the American semiconductor industry is the Semiconductor Manufacturing Technology Initiative (SEMATECH). Formed in 1987 with a five-year agenda to develop semiconductor manufacturing technology, the government contributes $100 million in matching funds to the fourteen member consortium, and in 1992 a new five-year agenda was adopted to ensure that the consortium maintain a role in the American industry as it approaches the next century.[6] The consortium represents the most significant partnership between industry and government to meet foreign challenges in an industry deemed critical to the United States.

The cultivation of government support for a private research cooperative was successful because it had a direct relation to defense. The Defense Science Board (DSB) reported in 1987 that it would be dangerous for national security if the Pentagon had to rely on foreign sources of state-of-the-art components. The report concluded that owing to Japanese competition and a wave of foreign acquisitions of American semiconductor materials and equipment firms, the

United States was in danger of losing the capacity to produce leading semiconductor manufacturing equipment.[7] Accordingly, the DSB reported that for the 25 semiconductor products and processes examined, Japanese companies led in 12 categories, American companies led in five, and relative parity existed in eight.[8] It was feared that failure to maintain this capability would place the entire semiconductor industry at a competitive disadvantage by forcing American companies to rely on foreign suppliers of manufacturing equipment. The DSB recommended that the Pentagon support and fund establishment of a U.S. Semiconductor Manufacturing Institute formed as a consortium of manufacturers The SEMATECH proposal passed as part of a broader defense bill, and in the following year Congress established the National Advisory Committee on Semiconductors to devise and promulgate a national semiconductor strategy as opponents of industrial policy deferred to promotion on the grounds of national security.

TRADE AND MARKETS

American technology policy has centered around market precepts. The Under Secretary of Technology of the Bush administration stated that "We are opposed to a 'top-down' approach to civilian technology development. The best way to support development of civilian technology is by providing incentives for innovation . . . [The Administration] believes in market-pull as opposed to technology-push."[9] The demand-pull approach isolates an important aspect of technological development, one that has been integral to the growth and prosperity of the Japanese industry. The fortunes of semiconductor device makers are inextricably linked to users who incorporate their products. Sales drive profits. Profits encourage capital expansion and investment. A diminishing demand base owing to a contraction of electronic sectors that utilize key technological components will affect that process.

The principal stimulus to semiconductor development comes from commercial applications. Military demand has not exceeded 20% since the 1960's, and throughout the 1980's and 1990's military demand has accounted for only 4 to 6% of overall semiconductor demand. The general trend is that electronic products and systems in consumer electronics, computers, and advanced electronic goods represent a shrinking market for American semiconductor producers. For example, the combined electronic systems produced in the United States and Europe consumed 63% of world semiconductors in 1984, but less than 47% by 1989.[10] Consumer electronics is dominated by Japanese producers; American world market share in computers has decreased from 54 to 38% between 1984 and 1989, and defense-related demand has diminished in an era of reduced military spending. The competitive disadvantages that the migration of semiconductor customers to Asia represents for the American semiconductor industry are compounded by limited access to the domestic markets serving those same electronic systems producers.

The European Single Market initiative has occasioned fears of a "Fortress Europe" and the semiconductor import regime would appear to bear this out. Although the EC reduced its tariff rates on semiconductors from 17 to 14% in 1985, those rates were not changed during the latest Uruguay round of GATT negotiations, and Europe now remains the only region in the world with protectionist semiconductor tariffs. By contrast, the United States and Japan have not had any significant tariff barriers on semiconductors for over 14 years, and none since 1985, when the two countries eliminated the residual 3% tariff on semiconductors. Yet American policymakers remain less concerned with European protection than they do with the apparently open Japanese market. This is because American semiconductor firms have been able to circumvent European import walls with foreign direct investment, and remain the leading supplier for the European market. By contrast, American and other foreign producers have been and continue to be marginal players in the Japanese market. Efforts to address this persistent condition have been central to American and Japanese trade disputes for years.

Fair and equal access to the Japanese market is not a trivial issue. The Japanese market was the largest in the world, accounting for 38.3% ($20.9 billion) of total world consumption prior to the recession in 1991.[11] The American market, ranked second in the same year with a total of $15.4 billion, has since rebounded on a surge of personal computer demand for semiconductors, claiming a market size of approximately $25 billion in 1993 compared to a Japanese market of about $24 billion.[12] Figure 5.7 illustrated the path of Japanese-American competition for world market share since 1982. According to one study, close to half of the decline in American market share can be attributed to faster growth in Japanese semiconductor consumption,[13] almost 90% of which was satisfied by Japanese producers. American firms have had immense difficulty in improving their market share to a level commensurate with Japanese import penetration in the American market or even much beyond 10% of total consumption. According to one simulation study of 16K DRAM chip production this implicit infant-industry protection was equivalent to a 26% tariff.[14]

Resistance to import penetration in the Japanese market has long been a thorny issue for the American industry. American pressure for liberalization of the Japanese market had promoted change, and by 1979 the trade regimes in the Japanese and the American semiconductor and computer industries were roughly comparable. However, in the aftermath of liberalization American companies remained frustrated in their attempts to increase their market position. At the same time, despite the superior technology and global market share of the American industry through the 1970's, there was a growing sense that the large, diversified Japanese companies posed a significant long-term threat. By the early 1980's technological advances in the Japanese semiconductor industry and alleged dumping by Japanese firms in the wake of the VLSI project made market access a more pressing issue for the American semiconductor industry.

While American firms were unable to penetrate the Japanese market, the Japanese share of the American market more then doubled in the 1980's, from 9 to 21%. These apparent asymmetries in competitive access and an eroding situation for American producers motivated the American government to depart from its traditional aversion to managed trade. In 1985 the Semiconductor Industry Association filed a trade petition under section 301 of the Trade Act of 1974, requesting the government to negotiate greater access to the Japanese market under the request of retaliation if access was not granted. At the same time, antidumping claims were filed against Japanese firms in the 64K DRAM and EPROM market.

These legal actions set in motion negotiations between the two governments that resulted in a comprehensive and enforceable trade settlement. Unlike earlier agreements,[15] the STA was more comprehensive, was more specific, and had formal legal status as an agreement pursuant to section 301(d) (2) of the 1974 Trade Act. President Reagan declared that failure to adhere to the arrangements of the accord would constitute a violation of section 301 and, therefore, be subject to punitive measures.[16] The agreement provided for the cessation of Japanese dumping and exacted a 20% commitment for American firms in the Japanese market before expiration of the agreement in 1991.

The persistence of Japanese dumping and efforts by some Japanese officials to downplay the 20% commitment led President Reagan to impose sanctions in April of 1987.[17] The demonstration of serious commitment and the use of sanctions to enforce that commitment induced Japanese compliance. Another motivating factor might have been the possibility that Japan would be unfavorably viewed under a tougher Super 301 provision emerging in the Omnibus Trade and Competitiveness Act of 1988. The change in attitude by Japanese officials was reflected in MITI's "buy American" promotion with the ten major electronic houses. American firms opened up 56 new sales offices, manufacturing facilities, design centers, and test/quality centers. Efforts were made to design American components into Japanese systems, and capital expenditures in Japan increased by 169% between 1986 and 1989.[18] Relations between respective trade associations became less confrontational, and they outlined a broad set of principles detailing foreign market access in Japan.[19] Konrad Seitz, head of the Planning Staff of Germany's Ministry of Foreign Affairs, observed that "the expansion of American semiconductor share in the Japanese market after the STA would not have occurred without the support of the American government."[20]

American market share in Japan rose from 8.5% in 1986 to 13.5% in 1991, and although this fell short of the 20% objective set before the termination of the agreement in 1992, the agreement had clearly benefited American producers. The American government, therefore, entered into trade negotiations with the Japanese government and signed a new five-year agreement, the New Semiconductor Agreement, that effectively extended the market access provisions of the old accord with some qualifications.[21] American firms continued a modest expansion into the Japanese market through 1990, but a leveling off at 13% the following year and the persistent gap in the level of imports and the 20% agreement remained a source of political tension between the United States and Japan through 1992. However, by early 1993 the 20% level was reached, albeit briefly, before foreign participation in the Japanese market dipped down to 18% for the year.

Figure 6.1 indicates that poor performance in the Japanese market stands in contrast to other world markets, where American companies compete with the same Japanese firms and generally hold an equal or leading market position. Other forces may be at work to limit access to the Japanese market. The structure of the Japanese industry is an important factor limiting foreign penetration. One reason may be found in the collusive arrangements (*keiretsu*) between Japanese semiconductor makers and the various electronic product producers that provide an expanded captive market. American semiconductor producers must compete with Japanese semiconductor producers that are typically a subdivision of the same electronic giant purchasing the chips and for markets that are to a greater or lesser extent within the orbit of *keiretsu* loyalty. The top eight semiconductor firms are divisions of electronic companies that account for 79% of Japanese semiconductor consumption.[22] The structure of the industry empowers semiconductor producers with the ability to influence the level of American sales in Japan. Most semiconductor consumption in Japan is done by Japanese firms with a distinct "buy Japanese" bias. Efforts to encourage electronic systems producers to purchase American semiconductors are thus more difficult than they otherwise might be.

Other reasons have been advanced to explain the consistently low American market share. One view is that low sales are a function of the market, one dominated by a demand for semiconductors in consumer electronics.[23] The United States has no consumer electronic industry capable of mounting a serious challenge to Japan, and this lack of user demand places American semiconductor manufacturers at a competitive disadvantage with Japanese firms, which have a long history of supplying industry needs. The Electronic Industries Association of Japan has argued that since American companies do not produce a lot of the type of semiconductors required by consumer electronic applications, there is a market mismatch and the United States cannot hope to achieve and maintain 20% of the Japanese market unless they gain a greater share within the market for consumer electronics. Moreover, since American companies have only a marginal position in DRAMs, the implication of the 20% commitment is that American firms must dominate important market segments of the logic market, including ASIC, microprocessors, and peripherals. However, it is unlikely that Japanese semiconductor producers will concede this high-end market to American producers. A 1987 Hitachi, Fujitsu, and Mitusbishi agreement to develop advanced 32-bit microprocessors with the objective of reducing Japan's dependency on American microprocessors (sold chiefly by Motorola and Intel)

Figure 6.1
Semiconductor Market Share, 1991 and 1993

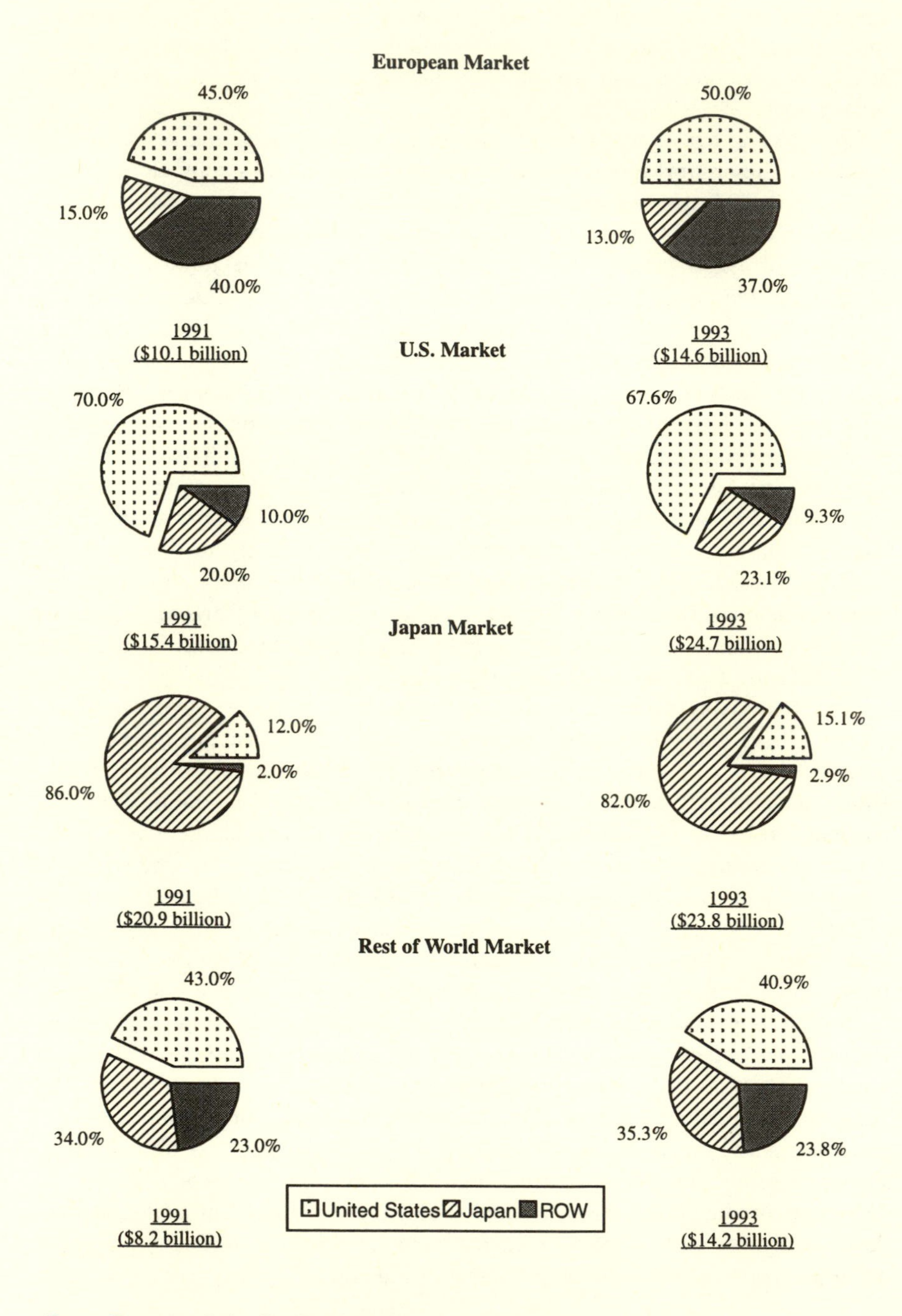

Source: Dataquest & the Semiconductor Industry Association

bears this out.

Another explanation is that the Japanese consider foreign semiconductors of poor quality. In the early 1980's this assertion had some merit. Japanese producers developed superior manufacturing skills and produced devices with a lower defect rate.[24] This derived from the targeting of DRAMs and concentrating production on major volume product, as compared to American companies which produced lower volumes across a wider array of custom products. However, by the mid-1980's the quality advantage in standard commodity products was no longer significant.[25] Transient market factors such as the over-valued dollar, and later, a depressed chip demand in Japan only partially account for the continued difficulty foreign manufacturers have confronted in increasing market share in Japan.

PENETRATING FOREIGN MARKETS

A largely insulated home market has provided economies of scale and a base from which Japan could challenge the United States for world leadership in the industry. Undercutting competition through export drives of inexpensive products has been an important strategy used to capture international markets, and in some cases has raised accusations of dumping. While this has not contravened GATT dumping regulations as determined by selling at less than fair value, the generation of excess capacity that can lead to dumping-like behavior is, in some measure, a by-product of the industry structure. MITI projects and consortium activities include the leading semiconductor producers, which operate in a highly competitive climate to market technological developments quickly and profitably ahead of the competition. This also fosters heavy investments in production capacity. The first to rapidly produce and market the chip can move down the learning curve and establish a leading position, but it also entails some risk because failure in the invested technology product or system can translate into significant losses. It is a condition that feeds a drive for overcapacity in the system as device makers race to the manufacturing gate. Whether or not these export patterns are predatory or competitive,[26] they had an enormous impact on the American semiconductor industry during the 1980's.

Large-scale systematic unloading occurred in 1981 and 1982 in 64K DRAM devices and in 1984 and 1985 in 256K DRAM devices. Both episodes were driven by falling semiconductor demand and rising inventories accrued from expanded production capacity. Japanese DRAMs were less expensive and offered superior quality to American versions. Enormous losses[27] were suffered by American and Japanese producers as the price of DRAMs fell precipitously. By the end of 1985 all but two American firms producing DRAMs were forced out of the market, leaving over 80% of the world market to Japanese producers. American firms have legal recourse by invoking antidumping regulations, but these measures are not effective deterrents to foreign dumping and they do not provide adequate safeguards for American firms considering investment in a product area where dumping is likely to occur. If injury relief is granted it can limit further damage but cannot repair material injury already incurred. Injured firms are not compensated for harmful dumping. Rather, future dumping is penalized. In the semiconductor industry and other high technology industries in which product life cycles are very short, a concerted wave of dumping can have negative consequences in less time than it takes the Commerce Department to investigate and prosecute an antidumping petition. Moreover, antidumping statutes are not applicable to third markets.

As a matter of strategy American companies were less capable of sustaining persistent losses. Notwithstanding the peculiarities of the American financial system, one ill-disposed to a long-term orientation, American semiconductor production is dominated by merchant houses that produce for sale in global markets. Japanese semiconductor production is dominated by diversified multibillion dollar multinationals in which semiconductors represent but one division. Given a strategic objective, Japanese producers have the financial staying power to sustain protracted losses to gain market share. Moreover, if all else fails Japanese firms have been able to rely on the cushion of a protected home market against the withdrawal or marginalization that confronts other producers.

The STA addressed the dumping problem in a more effective fashion. The agreement compelled producers to sell components that reflect production costs,[28] applied the pricing rules to all world markets, and set up machinery to detect violations and a mechanism for rapid retaliatory response in selected product areas. The end of DRAM dumping brought much needed stability back into the industry, and this fostered increases in capital spending and investment levels. Yet despite the new commitment to antidumping enforcement, the effects of past dumping persist. Japanese dumping was a catalyst that drove American competition out of this important industry, and high re-entry costs and the fear of future dumping have marginalized the American DRAM industry.

TECHNOLOGY

Innovation begins with capital investment in precompetitive technological development. Precompetitive technology is not necessarily product specific, but is engineered to create and develop the enabling tools and technologies that will be employed for new devices. The expense of new technology generation is rising owing to the increasing sophistication and complexities of the technology. For example, the process development costs for a single generation are over $150 million and rapidly increasing. Because of short product life cycles, major product generations must be introduced every three years, and a major process generation must be started five years before a product is brought to market. This is a financial burden that stretches the resources of even the largest firms, and it makes it more difficult for firms to go it alone.

In the past, American semiconductor firms did not engage in collaborative research as companies regarded precompetitive technology as proprietary. Consequently, the burden of extending all avenues of precompetitive technology fell on each firm, increasing the likelihood of redundancy, inefficiency, and a misallocation of resources that only a dominant industry could afford to endure. By contrast, company collaboration in precompetitive technological development lowers direct research costs and spreads risk, thereby reducing the investment risk attributable to any single firm. The efficiency of technology generation through the pooling of technological resources has long been understood in the Japanese semiconductor industry, and is slowly becoming an operational necessity in the United States semiconductor industry.

The long-lived hegemony of American microelectronics seemed to affirm the superiority of the American system. Until the 1980's the American semiconductor industry held a considerable technological lead in every major device and process technology. By the time Japan displaced the United States as leader in global market share in 1985 however, there was growing anxiety over the declining position of the American technology base. The 1987 Defense Science Board report indicated that the United States was behind Japan in a number of critical device, materials, and equipment technologies. An Office of Technology Assessment survey of semiconductor technology competitiveness places Japan ahead in ten of 17 technologies, the United States ahead in four, and the two countries even in the remaining three.[29] The National Advisory Committee on Semiconductors stated that "the United States was losing ground to Japan in key areas, holding ground in only a few areas, and gaining ground nowhere."[30]

Japan and America both enjoy leads in certain product subsectors. The Japanese have maintained a commanding lead in more standardized commodity products such as DRAMs. The United States is ahead in more technology-intensive products such as microcomponents, where manufacturing advantages are a less important determinant in competition. A lead has also been maintained in ASICs, and a strong position in EPROMs. However, the foregoing technology trends suggest that Japanese producers are on the way to challenging American companies in even the most advanced devices.

Over the past decade the Japanese semiconductor industry has invested more than its American counterpart in R&D, and this remains a decisive element in technological performance. A corollary to this has been the capacity of the government to single out specific technologies for development. In the 1970's and early 1980's those technologies were typically ones developed to catch up to and match established American capabilities. Since then, however, the government has focused on next-generation technologies that have the potential of revolutionizing the entire information field. By most accounts, the United States has fallen seriously behind Japan across a broad spectrum of next-generation technologies.[31] A major concern for the American industry is that the R&D initiatives undertaken in Japan will have a decisive impact on the competitive environment by the late-1990's. The Semiconductor Industry Association, for example, points to three specific next-generation technologies (synchrotron X-ray lithography, Josephson devices, and optoelectronics) that have the potential of producing breakthrough-type developments, and reflect the different ways the two systems pursue and develop microelectronic technology. Almost all the themes articulated in next-generation technologies are under financed in the United States and lag in the developmental efforts of Japan.

X-ray lithography may be the successor technology to optical lithography, the process by which the design of the semiconductor circuit is etched into the silicon material. The etching process is a critical aspect of semiconductor manufacturing. Most of the X-ray developmental efforts have been organized and supported by the government and associated public bodies such as NTT and the Japan Development Corporation. Public support has been channeled through use of the Photon Factory, NTT's synchrotron facility, MITI's Electrotechnical Laboratory, and development of the Spring-8 facility. In addition, SORTEC, a 13-member research consortium, has been organized with 70% of the $90 million budget provided by the Japan Key Technology Center. The American effort, by contrast, is underfunded and generally lacks strategic direction. Japan has seven operational synchrotron facilities and the United States only one, at IBM. The American government has provided spotted, though increasing, support through DARPA and the Department of Energy, but together the American resources do not match Japanese efforts. Promising research avenues do not always materialize even when resources are targeted, but should a particular technology live up to or exceed expectations, the organization that is furthest down the learning curve stands the best opportunity to reap the greatest reward.

Josephson devices represent another revolutionary technology. These superfast electronic switching devices have the potential to give desktop computers the speed and power of a supercomputer. The Japanese effort in this area is integrally tied to a long-term objective of creating ultra-high-speed computers. The VLSI project (1975-1979) began work on Josephson technology, and one-third of the resources spent on MITI's Supercomputer Project (1981-1989) went toward Josephson technology; by the end of the 1980's Japanese research efforts were paying dividends as prototypes of memory and logic devices were produced. In 1989, the ERATO (Exploratory Research for Advanced Technology Organization) program commenced funding on a five-year project to develop integrated circuits to improve existing Josephson technology. And in 1990 MITI began construction of a Josephson computer at its electrotechnical laboratory.

Josephson technology was pioneered by IBM during a 12-year program financed in part by the National Security Agency. That program ended in 1983 when it was believed that the research would yield benefits only marginally superior to present silicon technology, in large measure because it was believed that Josephson technology was not applicable at the submicron level. Since 1983, there has been no significant effort in the United States to develop this technology. Indeed, Japanese research in Josephson technology may have gone the way of IBM had it not been for the financing role of the Japanese government. A member of the Supercomputer Project commented that "MITI did not allow the research to stop."[32] In recognition that related technological advances have given Josephson technology realistic commercial potential, a group of American companies and national laboratories have sought to organize an American response in a six-year, $113 million effort in which the government would fund 60% of project costs.[33]

Optoelectronics includes high-speed computing, optical communications, compact disks, and a variety of related technologies. It is expected that optical integrated circuits will be commercialized by the mid-1990's as an alternative to silicon-based semiconductor technologies. That Japan has an optoelectronics industry can be attributed to systematic government initiatives to create one. Between 1979 and 1987 MITI organized the first optoelectronic project centered around a joint laboratory and $75 million in government funding. In 1986 a second-generation optoelectronic project was launched with $83 million from the newly formed Japan Key Technology Center and $29 million in equipment from MITI. In 1989, MITI launched a third optoelectronic project with a ten-year horizon. In 1980 Japan had no such industry, but by 1991 its industry had revenues in excess of $27 billion and dominated the world market for that technology.[34]

Not all advances in optoelectronics have been by-products of government-sponsored R&D projects, but the process has greatly accelerated Japan's drive into the upper technological echelon of the industry. MITI accelerated technology development, signaled that this was a sector to be developed and thereby reduced the investment risk by private firms, and ensured the dissemination and adoption of new optoelectronic technologies broadly across the Japanese electronics industries. American firms still enjoy a strong position in optical systems, but are in jeopardy in the area of commercial optical components. A number of surveys over the past five years indicate that the United States is behind and falling further behind in this technology.[35] The U.S. International Trade Commission stated that "the Japanese are so far down their learning curve compared with the U.S. that they represent a considerable threat to the U.S. industry in that which is fast becoming a cost-driven, commodity business."[36]

The case of the above next-generation technologies is indicative of what has been occurring more generally in semiconductors. MITI has a diminishing role, but as Table 6.1 suggests, there are many microelectronic themes promoted by MITI and its satellite organizations. More generally, Table 6.2 indicates that European and Japanese firms operate in an institutional framework that eases the burden of capital formation for microelectronic R&D. MITI can augment the availability of capital in selected critical areas, but the vast bulk of R&D is conducted by Japanese semiconductor producers. These Japanese companies continue to pursue technologies across a broad front and to outspend the United

Table 6.1
Microelectronic Research Sponsored by MITI in the 1990's

Sponsor	Project	Purpose
NEDO	Superlattice Devices	Device technology
NEDO	Three Dimensional IC's	High density IC's
NEDO	Bioelectronic Devices	Biodevices on human information processing
NEDO	Nonlinear Photonics	Optical signals in IC's
MITI	Real World Computer Project	Sixth Generation Computer project. Optical, neural, and parallel computing
Key Technology Center	Sortec	X-Ray lithography
Key Technology Center	Optoelectronic IC's	Optoelectronic and optical IC
MITI ETL	Josephson IC's	Device and packaging technology
MITI ETL	Nanomechanism	IC's created through micro-manipulation
NEDO	Superconducting devices and materials	High temperature superconducting
NEDO	Ion Engineering	Ion beam lithography
NEDO	Applied Laser Engineering Center	Excimer laser technology

Source: Adapted from Thomas R. Howell, Brent L. Bartlett, and Warren Davis, *Creating Advantage: Semiconductors and Government Industrial Policy in the 1990's* (Semiconductor Industry Association, 1992).

Table 6.2
Capital Investment and Investment-Related Policies in Microelectronics

	United States	Europe	Japan
Direct Capital Subsidies	NO	YES	NO
Subsidies for Operating Losses	NO	YES	NO
High Technology Development Banks	NO	YES	YES
Bank Ownership of Strategic Industries	NO	YES	YES
Government Allocation of Credit	NO	NO	YES
Special Depreciation Allowances	NO	NO	YES
Government Ownership of firms	NO	YES	NO
Support from state and local government	YES	YES	YES
Technology specific tax breaks	NO	YES	YES

Source: Adapted from Thomas R. Howell, Brent L. Bartlett, and Warren Davis, *Creating Advantage: Semiconductors and Government Industrial Policy in the 1990's* (Semiconductor Industry Association, 1992).

Figure 6.2
Semiconductor R&D Spending 1984-1993

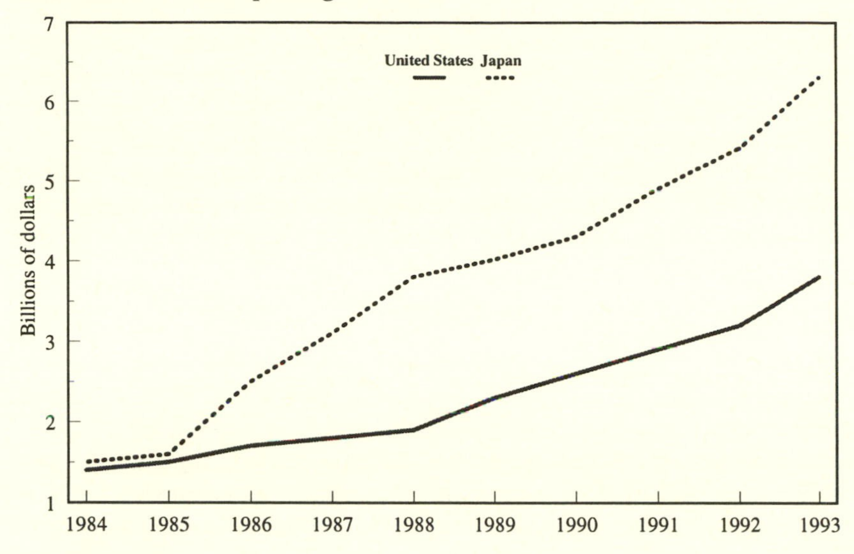

Source: Dataquest

Figure 6.3
Semiconductor Capital Expenditures 1982-1993

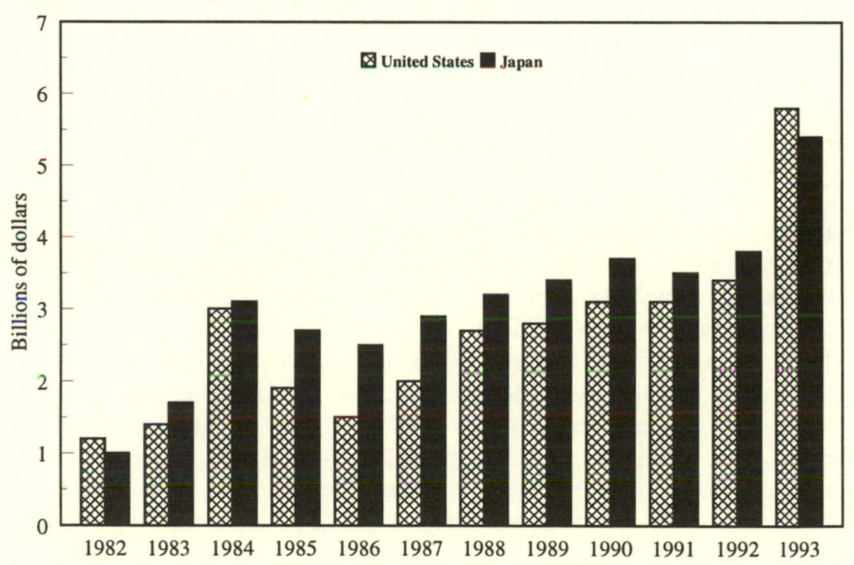

Source: Dataquest

States industry in research and development. Figures 6.2 and 6.3 illustrate the trend in R&D spending and total capital expenditure on the company-level. Since the American industry has been unable to match those resources, it is obliged to be more selective in what technologies to pursue. Japanese investing in a broad array of potential technologies may prove costly, but it positions them to reap the commercial windfall of new technologies. The Semiconductor Industry Association stated: "it is evident from the scale and scope and intensity of Japanese efforts that Japan will make at least some major technological breakthroughs during the next decade, and, because of its industrial organization, is in excellent position to exploit those breakthroughs in the commercial arena.[37]

INFRASTRUCTURE

The semiconductor industry consists of two mutually dependent sectors, the semiconductor device manufacturers and semiconductor material and equipment (SM&E) firms. The SM&E firms provide the manufacturing equipment and raw materials that manufacturers transform into semiconductor devices, and the ability of manufacturers to produce semiconductors efficiently depends on timely access to state-of-the-art materials and equipment. The United States dominated these areas of the semiconductor industry as it once did devices,[38] but in the 1980's the American SM&E industry suffered as a result of Japanese competition. World market share for American SM&E firms decreased from 76 to 44% from 1981 to 1990 and, during the same period, increased for Japanese SM&E firms from 21 to 49%.[39] Accordingly, in 1980 the top ten SM&E firms in the world were all American, but by 1990, four of the top five and six of the top ten were Japanese. By 1990 Japanese firms had effectively captured world leadership in sales and had also assumed exclusive supply positions in key sectors.[40] Consequently, American semiconductor manufacturing firms have become more dependent on foreign sources of materials and equipment.

The industry network of American SM&E firms is highly fragmented, dominated by hundreds of small firms, most of which have annual sales of less than $20 million.[41] As a result, the American SM&E industry suffers several structural disadvantages compared to its larger Japanese counterparts. First, the small SM&E companies are hard-pressed to meet the demands of escalating capital requirements and R&D costs. Second, their small size makes them more vulnerable to acquisitions by foreign conglomerates. Third, in cases where these firms have state-of-the-art technology, the challenge of sustaining a competitive position over the long-term is aggravated by the shrinkage of the domestic demand base and the gradual migration of their markets to Asia.[42] Semiconductor production and capital expenditures data from 1984 to 1990 indicate that every $5 of semiconductor production creates $1 of SM&E demand.[43] In this respect, the virtual loss of the DRAM market compounded the problem because equipment and materials are usually first developed for memory production. The expansion of the Japanese semiconductor device industry has had a pull effect on the Japanese SM&E industry, while the drag of a deteriorating American chip industry has been manifested in a decreasing SM&E world market share for American firms.

The SM&E industry is composed of four major segments. The semiconductor equipment portion represents three of those segments (accounting for most of the capital equipment purchases in the semiconductor industry): wafer fabrication equipment, assembly and packaging equipment, and test equipment. The fourth segment is processing and packaging materials. The American market share in SM&E sectors such as semiconductor materials, optical lithography, and automated testing equipment has been dramatically reduced.[44] These three equipment markets are important in the chip production process and their eclipse is emblematic of a general retreat by American firms in many equipment markets.

Japanese efforts in semiconductor materials gained momentum in 1984 when MITI outlined goals to position Japan to be the leader in high technology business areas that depend on advanced materials. Government directly funded research and development of advanced materials, and organized specific industry workshops to facilitate plans for expansion, investment, and joint R&D projects. Optical lithography (subsector of wafer fabrication segment) and automated testing equipment are two other SM&E sectors that have benefited from government intervention. The leading Japanese producers of optical lithography, Nikon and Canon, trace their rise in the industry to the VLSI project in the late 1970's. This provided a catalyst to develop new equipment technology with a ready market and forged strong vertical linkages with semiconductor device makers.[45] In automated testing equipment, Advantest, the world's largest producer, owes its success to direct government intervention.[46] The signaling of this sector for development and laws enacted to subsidize costs for the development of materials technologies encouraged many Japanese firms to enter the field, a development that hastened the exit of American firms from the market.[47]

Access to state-of-the-art manufacturing equipment and materials is essential to remaining competitive in semiconductor devices. A 1990 study by the Defense Science Board revealed that one of only two types of Japanese suppliers of a specific semiconductor manufacturing equipment withheld the most advanced models from foreign customers for two years after they were made available to Japanese customers.[48] In the United States, growing recognition that self-sufficiency in the tools and materials necessary for semiconductor manufacture had been lost, the fear of strategic withholding, and the national security ramifications of semiconductor dependency, together motivated a joint industry and government response. The SEMATECH initiative was designed to reduce this dependency.

SEMATECH is focused on semiconductor manufacturing technology. The overriding objective is twofold: to reduce the dependency on Japanese manufacturing equipment and materials; and to enable American semiconductor device manufacturers to regain world leadership.[49] SEMATECH identified a research agenda and, in so doing, provided an important sense of direction in the SM&E industry. Although SEMATECH became virtually the sole source for long-term R&D in the industry, the seriousness of the competitive deficit forced the consortium to be more vertically integrated (beyond generic technology research) by providing for a near—and medium—term manufacturing research agenda as well. The consortium has constructed a state-of-the-art semiconductor manufacturing plant; holds conferences and workshops to establish targets and standards and evaluate improvements in manufacturing technology; and has outlined eight formal objectives.

An important aspect of the SEMATECH initiative was granting contracts to American SM&E producers to develop equipment, materials, and systems that would be tested and implemented at the SEMATECH pilot facility in Texas. These development contracts assisted the industry by covering a portion of development costs; facilitated the use, testing, and improvement of that equipment to make it more readily acceptable for industry employment; encouraged joint development among groups of SM&E firms, and the adoption of uniform standards. Development contracts have farmed out specific R&D schemes that complement other efforts to enhance SM&E technology. Research has progressed through collaboration, not duplication.

An important contribution of SEMATECH has been to foster closer working relations between SM&E suppliers and semiconductor producers.[50] According to SM&E executives, ineffective relationships between semiconductor manufacturers and their suppliers was a close second to the cost of capital as the most significant factor compromising their ability to compete effectively. The same executives considered the ultimate success of the consortium to be contingent upon the extent to which SEMATECH fosters effective and long-term relationships between semiconductor manufacturers and American suppliers.[51]

Noting that Japanese suppliers have benefited from close interactions with customer firms, SM&E executives cited the need to share technical performance data for equipment with suppliers, to involve the suppliers in the planning process of chip generation, and to have device makers share some of the costs and risks associated with developing the next-generations of semiconductor fabrication equipment and materials.

In the past, semiconductor manufacturers have been reluctant to permit equipment suppliers to thoroughly examine how the equipment operated in the manufacturing environment because the device maker could incorporate incremental improvements into SM&E systems that could benefit rival semiconductor producers. This compromised SM&E advances by denying them an important test laboratory. SEMATECH has attempted to strengthen inter-industry relations by having consortium members share strategic goals, technical information, and some of the cost of new equipment generation with suppliers. In addition, the consortium built a fabrication facility in Texas to provide a laboratory where new devices and equipment could be tested and refined. The Texas plant provides valuable feedback between users and suppliers. SEMATECH established a chapter for 140 semiconductor equipment and materials producers (SEMI/SEMATECH), and arranged regular meetings for equipment producers that made losing bids for SEMATECH contracts, highlighting areas that required improvement.

Determining the overall impact of any large and complex research effort is a difficult and multifaceted task. However, there is some preliminary evidence that SEMATECH has had a broad-ranging beneficial impact on the American semiconductor industry. For example, America's two largest semiconductor producers, National Semiconductor and Intel, created manufacturing facilities in 1991 that largely employed American-made machines. Motorola's MOS-11 fabrication facility featured 75% foreign-made equipment prior to SEMATECH; by 1991 80% of the equipment at this facility was American-made. Similarly, Intel purchased $150 million worth of American equipment and materials in 1991 that, without SEMATECH, would have been procured from foreign sources.[52] SEMATECH has also improved systemic flaws in the industry. The industry's structure has become more rational with the adoption of uniform standards, joint-development R&D, and the refinement of industry products in the consortium testing facility. The consortium has driven standards for equipment safety, cleanliness, and systems integration, permitting members to reduce the time and cost to evaluate and debug machines that have been approved by Sematech.[53] Even though SEMATECH has had to narrow its choices on which of the hundreds of manufacturing steps to focus on, the process of conducting industry workshops has provided market information of some value to firms with only limited access to capital. Some firms could postpone or concentrate R&D efforts because SEMATECH had discovered that current technology will suffice in one or another aspect.[54] SEMATECH has made relations between suppliers and manufacturers less adversarial and more cooperative. Although the results have benefited larger members over the smaller members and nonmembers, the consortium has contributed to an improving competitive performance by American semiconductor producers.

In 1990 the U.S. National Advisory Committee on Semiconductors reported that despite a growing worldwide market for SM&E, the American industry was projected to lose worldwide market share at the rate of 2% per year.[55] SEMATECH has helped reverse the erosion of American SM&E market share, and by 1991 improved performance of American SM&E was reflected in the recapture of 3% of world market share. By the end of 1992, American SM&Es had increased their worldwide market share to 53%, while the Japanese industry lost further ground, falling to 38%.[56]

This reversal of fortune can also be attributed, at least in some measure, to shifts in capital spending. In the United States, capital spending by chip makers for SM&E products increased 19% from 1991 to 1992, in part to accommodate the growing demand for PC-relevant semiconductors, while Japanese capital spending decreased 39% during the same period owing to the persistent economic downturn at home. Since each country relies predominately upon its own equipment makers, this was clearly a factor in the gain in American market share. The correlation was further demonstrated by the slight dip in American market share to 48%, when capital spending by Japanese firms actually rose by 30%.[57] Since part of the gain in American worldwide SM&E market share can be attributed to immediate demand conditions in both economies, at this point it is difficult to project the long-term impact of SEMATECH. One study by VLSI Research Inc. did indicate that "underlying strengths suggest it will last beyond the Japanese slowdown."[58]

The consortium has arguably contributed to the short-term rebound of the American industry, but it should not be regarded as a panacea because its impact on long-term competitiveness in the industry remains to be determined and many problems remain beyond its scope. First, SEMATECH lacks the resources to commit equal efforts to the struggling materials industries and has not, therefore, been able to enhance the competitive position of this important sector. Second, SEMATECH cannot prevent the transfer to foreign ownership of significant portions of the American SM&E industry.[59] And third, it cannot substitute for a declining demand-pull as more semiconductor production is being conducted outside of the United States by Japanese device makers with a preference for Japanese SM&E. According to senior executives from 31 equipment and materials suppliers surveyed by the GAO, the high cost of capital, under-developed relationships among the consortium's member companies and their suppliers, and other factors that led to the decline of American SM&Es still confront the industry and seriously constrain any long-term efforts to strengthen it.[60]

CAPITAL

The ability of a firm to gather sufficient funds for capital investment, either through retained earnings, stock issues, or the assumption of debt, is of primary importance. The National Advisory Committee on Semiconductors commented that "the single most important consideration for the current and future health of the semiconductor industry is the availability, cost, and patience of capital."[61] The level of capital investment in the industry is influenced by two fundamental issues: the cost of capital and market performance. A contracting sales base translates into lower revenues, which in turn leads to a smaller pool of retained earnings with which to finance capital investment. On the debt side, poor performance increases the cost of borrowed capital because of additional risk. A troubled industry, therefore, may suffer a capital shortfall, which in turn has a compounding effect on competitiveness.

This has been the recent fate of the American semiconductor industry. Despite high levels of investment relative to other American industries,[62] the

Japanese industry has outspent the United States in semiconductor plant, equipment, and R&D by $12 billion from 1984 to 1991. This trend was reversed in 1991 to 1993, as capital spending by American firms exceeded similar spending by Japanese companies, but that likely reflects an episodic short-term gain with U.S. chip makers driven by pent-up computer semiconductor demand, while the Japanese economy was suffering its worst recession since 1945. Capital investment is a vital concern because investment in new processing equipment and research and development are the "lifeblood of productivity gains and technological improvement."[63] Without a high level of investment, the U.S. semiconductor industry cannot remain competitive with the best in the world in technology or product costs.

A major concern for American companies is that Japanese producers may not be constrained by similar limitations. An important advantage for Japanese companies has been the availability of abundant low-cost capital, especially at crucial junctures in semiconductor development. The collapse of the Japanese bubble economy, which was fueled by overvalued stock prices and real estate prices, has certainly limited the availability and increased the cost of capital to Japanese chip makers.[64] Although the cost of capital in Japan is higher than it was several years ago, firms operate in a more benign interest rate environment than is found in the United States, and in an industrial structure that reduces risks that would otherwise increase the cost and availability of capital for investment purposes. For example, Japanese development banks continue to extend low-interest loans for selected high technology projects. Other important factors influencing the cost and availability of capital have been the *keiretsu* and the large size and product diversification of Japanese semiconductor producers, a condition that diminishes financial risk.

The Japanese economic system is more conducive to the strategic management of capital. Because Japanese companies rely more on debt financing in the form of direct bank loans,[65] the *keiretsu* structure is in a position to facilitate the flow of capital to growth areas signaled by MITI initiatives. This structure facilitates the flow of capital to strategic sectors, reduces the possibility of a capital shortage, and contributes to the "patience" of capital by tying the interests of the targeted industry with fellow *keiretsu* members, thus reducing the probability of capital flight. Investment theory holds that capital gravitates to where the reward is greatest. American firms are more reliant on equity financing, and with this comes a greater awareness of investor sensibilities. Higher risk requires a higher rate of return. Attracting capital depends on offering a higher return than could be achieved by the investor elsewhere given a similar level of risk. The declining fortunes of the semiconductor industry in the 1980's increased the risk for individual investors and compromised the ability of the industry to generate an adequate return on investment. Because capital raised today on financial markets can migrate with few constraints tomorrow, managers must think in terms of short-term performance to avoid capital flight. The rebound of the American industry since 1991 has made it an attractive investment opportunity for those same money managers, and many semiconductor stocks have had a significant role in pushing the nation's stock market to record levels. At the same time, this orientation discourages the type of long-term investment perspective needed for success in the industry.

Declining sales performance hinders access to fresh capital and generates higher risk, adding to what is generally considered an already higher cost of capital in the United States. A 1991 General Accounting Office report surveyed 31 leading SM&E firms and found that senior executives regarded the high cost of capital as the leading factor affecting declining fortunes of the suppliers industry.[66] This has influenced the low levels of investment in plant and equipment by American SM&E firms compared to the level of investment by Japanese suppliers. A lower cost of capital allows for greater debt capacity against existing assets. It also reduces risk in investment decisions by reducing the required rate of return thus facilitating more investment, more investment with longer time horizons, and a greater immunity from cyclical swings in demand.

A related factor influencing capital formation is the extent to which investment risk may be managed. The life cycle of semiconductors is short—two to three years—so that an investment must be recouped within a limited period before reinvestment in the next process or product technology. The cost of staying at the leading edge has increased sharply. Since 1980, the cost of an average state-of-the-art facility has risen 800%. Also, to build a production facility from scratch will cost anywhere from $400 to $800 million, an investment that cannot be adjusted if sales projections fail to meet expectations. Semiconductor investment tends to be risky and expensive for any nation, but how that risk is managed may diminish or compound the cost, and thereby the level and frequency of investment. Short product life cycles coupled with large front-end investment necessitates a rapid return on investment. Assured access to large markets provides a competitive advantage in sales volume and economies of scale that can minimize that investment risk. As noted earlier, Japanese producers effectively own their domestic market for semiconductors, the largest in the world. American companies by contrast face more competition in the local market, remain vulnerable to dumping, and enjoy only token access to the Japanese market. These differences in market structure increase investment risks for American firms and reduce attendant risk for Japanese companies.

Another way that investment risk is reduced in Japan is through sharing that risk with government. Risk may be lessened by creating development banks for high technology or by having the government as an equity partner in the venture. All the principal competitors of the United States have one or more government banks funding high-risk investments in semiconductors and related high technology enterprises.[67] The American government has avoided assuming an equity stake in an enterprise. Accordingly, a recommendation by the National Advisory Committee on Semiconductors to create a Consumer Electronics Capital Corporation to provide low-cost "patient" capital to shore up the electronics industry was rejected on the grounds that it smacked of industrial targeting.[68] In Japan, government makes equity investments in research companies for costly technologies that might otherwise suffer underinvestment, or to shore up enterprises deemed critical by the state.[69] This financial advantage is magnified as it attracts private capital into the sector by signaling that it is an industry that will not be permitted to fail.

Tax policy has a direct bearing on capital formation in the semiconductor industry and has been an important determinant in the relative discrepancy of capital investment between Japan and the United States. The United States does not have specific tax provisions designed to promote investment in the semiconductor industry. Generic tax provisions such as the Research and Development Tax Credit are beneficial to semiconductor producers, but a 50% tax on capital gains has had a very negative impact on semiconductor investment.[70] Japan and Germany provide for generic R&D tax credits and do not tax capital gains. Furthermore, Japanese tax policies complement generic R&D tax credits applicable to all industries with specific tax measures designed to promote investment in the semiconductor industry. Such measures include an accelerated depreciation on semiconductor equipment[71] and application of the Key Technologies Tax Credit to all semiconductor technology. The Key Technologies Tax Credit goes so far as to list specific pieces of semiconductor equipment. The absence of a capital gains tax and specific promotion of the semiconductor industry through tax credits are other reasons why Japan has been able to outspend its American counterpart.

The above examination has simplified more vast and complicated differences in the two financial systems. Nonetheless, it does illustrate the problem of capital formation in an intensely competitive industry and the strain that high capital costs place on investment in plant, equipment, and R&D. The tendency for American managers and private investors to redirect their funds to areas that have lower risk and promise higher returns is sound financial management, but potentially dangerous from a national perspective. Not all systemic differences affecting patterns of capital formation are attributable to government intervention, but together they continue to represent a source of strength for the Japanese semiconductor industry.

CONCLUSION

It is not enough for a government to announce the intention to create a world-class semiconductor industry. For Europe it remains an elusive goal despite years of government promotion. The global success of the Japanese semiconductor industry since 1970 can be attributed to a series of factors. Several factors are part of a well-conceived state strategy. Other factors are not. Although the level of subsidy attached to that strategy has been relatively small, government intervention has been pervasive and continues to exercise a great deal of influence in building consensus and in maximizing the deployment of industry resources. At the same time, state efforts would not have succeeded without the investments, strategies, and structural strengths of Japanese firms. It is reasonable to conclude that public policy has facilitated and accelerated the development of the Japanese semiconductor industry, and equally reasonable to suggest that the Japanese industry would have gradually evolved as a world-class industry even in the absence of sector-specific government policy.

Industry-government consortia have been an important element in the technology infrastructure of Japan. However, it is the degree of effective use of this infrastructure, along with other factors, that has ultimately determined competitive position. The Advisory Council on Federal Participation in SEMATECH stated that "even at their most successful, SEMATECH and similar measures are palliatives—selective and temporary efforts to compensate for general conditions in the U.S. economy that have contributed to competitive weakness.[72] SEMATECH cannot turn the American semiconductor industry around by itself. European efforts in JESSI and ESPRIT have been criticized for having produced little commercially useful technology, but part of the problem is that the process has been compromised by other conditions that include an inferior technology base, a highly segmented industry structure, and diminishing demand-pull that make even promising technologies difficult to commercialize successfully. SEMATECH has enjoyed some short-term success because of a surge of demand in downstream markets. It has focused on an obvious need to rationalize industry standards and relationships, while members have focused on research problems common to all members, and elements of urgency associated with national security considerations appear to have been a motivating force.[73] However, long-term problems remain. As the varied strategies and experiences of the United States, Europe, and Japan have illustrated, the overall utility of government-industry cooperative research also depends upon related factors influencing competitive position.

The level and intensity of government promotion have passed through two distinct phases since 1970. During the first phase, between 1970 and 1980, government involvement was extensive and critical. The stimulation of user industries such as computers and telecommunications, R&D subsidies, and an insulated home market built up manufacturing advantages and developed competitive strengths that were translated into gains on the international market. R&D subsidies through research projects such as the VLSI project provided an important source of capital for firms that, at that time, could not match the R&D investments of American companies. And, without protection, the Japanese industry would have had a difficult time competing with the cheaper state-of-the-art American components. American firms would not have been as ready to license technology, and instead would have probably occupied a lucrative supply position in Japan. Superior technology, large-scale production facilities, and first-mover advantages could have placed American firms in a dominant position equal to the one they enjoyed in Europe.

Public policy was most influential during the catch up period. Once the semiconductor industry matured in technology and market share, the Japanese government became less directly involved and has since led by what one might term soft policies. Now that Japan is at the technological frontier there is no industry leader to emulate, no large pools of technology to license, and no obvious need to pursue an import substitution program. That does not mean that government has lost its capacity to influence the industry. The government encourages R&D through the tax policy, loans and subsidies, and government industrial research labs. The cooperative R&D ventures promoted by MITI to develop next-generation technologies institutionalizes the drive to extend the technological frontier. This is an important role and it suggests that industrial policy is not simply limited to a period of catchup. These efforts not only rationalize the R&D agenda but also have an important signaling effect that attracts economic resources into that activity. MITI remains an important instrument in consensus building, and a vehicle for the exchange of information and ideas, but its capacity to manipulate comparative advantage in the industry has lessened considerably in the last ten years.

From the perspective of international trade however, even if the semiconductor industry no longer benefits from government intervention, that it once did constitutes "original sin." The argument among several prominent American observers is that although the overt trade and promotional industrial policies are less obvious than they were ten years ago, present competitive advantage was conferred by government intervention.[74] Import restrictions were dropped in 1976, and through the 1980's and into the 1990's the Japanese government has had no policy to restrict foreign market share in Japan. Rather, the opposite has occurred as MITI has attempted to raise the level of American market share in Japan. However, the difficulty in penetrating the Japanese market is, in some measure, a residue of government intervention in the 1960's and 1970's. Tight government control and supervision over all aspects of technology development and investment fostered an economic structure that has made American participation less than optimal. The issue in the 1990's is that more transparent structural and market-related barriers have replaced overt home-market protection.

The market access issue is the central feature in Japanese-American semiconductor competition precisely because the closed Japanese market confers an unfair advantage for Japanese producers. Japanese world market share gains in SM&E and semiconductor devices are partly explained by the rapid growth of the Japanese semiconductor market. User demand has always been an important force driving innovation, in achieving economies of scale, and in generating sales necessary for future capital investment. In this context, the drive to open the Japanese market assumes strategic significance. American efforts to regain the initiative in the industry will be compromised as long as Japanese producers continue to dominate in such a large market.

The STA has made managed trade an operational reality and in so doing has codified two principles. First, that predatory business practices, whether the result of industry structure or of government policy, will not be tolerated and American industry will be protected accordingly. Second, that Japan must open its market if it wishes to continue to sell to the rest of the world. Since 1987, the United States has not threatened to impose retaliatory measures in light of unsatisfactory market access, but the issue has remained a source of friction between the two countries. Nevertheless, efforts to open the Japanese semiconductor market through managed trade have had some success. American market share has increased under the STA, and it has offered the best opportunity for American semiconductor producers to expand their presence in an important industrial market.

American market share in all other significant markets is much higher. In Europe, American firms lead 45% to 15% for Japanese companies. In the rest of world market (not including United States, Europe, Japan) the lead is 41 to 35%, and in the United States, American firms maintain a comfortable lead by a margin of 68 to 23%. (see Figure 6.1). American companies have by no means become marginal players. On the contrary, it would suggest that the Japanese lead in overall market share stems from 86% control of the domestic market, the fastest-growing semiconductor market in the world and one 25% larger than the American market for semiconductors. In addition, recent trends in global competition indicate the American industry is getting stronger. American firms have increased their world market share in each of the last two years, and during that same period Japanese world market share has dropped. In 1993, the American semiconductor industry reassumed the position as the leader in world market share by the slim margin of 43 to 41%, and retained that lead in 1994 by a margin of 41 to 40% for Japanese companies.[75] It is the first time that an American industry has recovered the lead in an industry after it has been lost to Japanese competitors. Part of this transformation must also consider the dramatic appreciation of the yen and depreciation of the dollar in recent years, a development that has made American semiconductor exports more attractive and Japanese shipments artificially expensive. The exact impact of these exchange rate movements are difficult to quantify but they most surely must be viewed as a corrollry factor in American semiconductor competitiveness. As Table 6.3 illustrates, American firms have gained in the global ranking as Motorola and Intel have pushed past Hitachi, and Texas Instruments passed Mitsubishi.

Improved performance has demonstrated that panic over the declining fortunes of the American semiconductor industry may have been premature and that the march of the Japanese juggernaut through the industry is not inexorable. The competitive resurgence is partly related to market opening measures in Japan, but more fundamentally it has been driven by a surge in domestic demand. The American chip market grew by 16% in 1992, while the Japanese market actually dropped 5%.[76] Whether this rebound reflects systemic strength or cyclical demand is open to debate. Japanese producers focused most on production for consumer electronics and in the computer market for mainframes, two markets

Table 6.3
Top Ten Semiconductor Vendors Worldwide
(millions of $)

Company	1989	Company	1994
NEC*	4,964	Intel	10,121
Toshiba*	4,889	NEC*	7,944
Hitachi*	3,930	Toshiba	7,527
Motorola	3,322	Motorola	7,237
Fujitsu*	2,941	Hitachi*	6,485
Texas Ins	2,787	Texas Ins	5,280
Mitsubishi*	2,629	Samsung	4,893
Intel	2,440	Fujitsu	3,858
Matusushita*	1,457	Mitsubishi*	3,735
Philips^	1,690	Philips^	2,905

*Japanese Firm
^ European Firm

Source: Dataquest

that are currently in decline. Consequently, the recession in Japan has hit the semiconductor industry particularly hard and slack demand for memories has accounted for most of the loss in Japanese market share.[77] An increased demand for microprocessors driven by strong personal computer sales in the United States accounted for most of the change in American market share. In addition, a corollary demand for more memory to handle the increased power was mostly picked up by American producers, which as the dominant suppliers to the domestic market (70%) were best positioned to take advantage of rising demand. However, American semiconductor companies should avoid relying upon a privileged position in the personal computer market because if that market declines, as was the case for consumer electronics in Japan, then the lack of diversity will make it difficult to retain market share and the momentum built up over the past two years.

Recent demand has buoyed the American industry and through relentless innovation by leading producers such as Intel American preeminence in the highest technology markets in the industry has been assured, at least through the mid-1990's. Nevertheless, Japanese producers will likely continue to pose a long-term challenge to American producers as several factors continue to shape certain advantages enjoyed by the Japanese semiconductor industry. First, the industrial milieu is such that the government, through relatively small gestures and modest research subsidies, can induce the commitment of financial and technological resources toward the attainment of a particular commercial objective. That investment has been given greater institutional clarity in the race to develop next-generation technologies. Second, the industrial structure provides defacto protection to the industry. Although a 20% commitment may be difficult to realize now that American production in more standard devices has been marginalized, the inability to break into the market despite liberalization suggests that other collusive forces are at work. Third, the gradual migration of downstream markets to Japan (and Asia more generally) has made the domestic market one of the largest semiconductor markets in the world. This provides important demand-pull advantages and growing economies of scale. Finally, the availability of relatively cheap and patient capital, due in part to a more benign macroeconomic environment and in part to industry and firm structure. It has been instrumental in the success of the industry and it permits Japan to outinvest the United States in many critical areas of R&D. Perhaps the single most important challenge facing the American industry is maintaining leading technological capabilities without long-term advantages in capital formation.

It is significant that these factors are not a function of industrial targeting because it indicates that other forces have shaped Japanese advantage in the industry. It is true that government has influenced capital formation through tax policies, R&D subsidies, and low-interest loans. On balance, however, the scale of funding has not been significant relative to the total investment requirements of the industry. The MITI R&D budget through the 1980's has averaged $750 million a year. This is dispersed throughout industry, 18% of which went to high technology projects, 11% to microelectronic-related R&D.[78] When considered against the billions of dollars invested in R&D annually by the top six Japanese semiconductor producers, it is clear that the fortunes of the industry are fundamentally determined by private-sector participants. Similarly, support from the Japanese government through cooperative research in microelectronics has been only one of several factors contributing to the strength of the Japanese SM&E industry. Other factors included the advantage of a growing domestic market, close working relationships between Japanese semiconductor manufacturers and their suppliers, the availability and lower cost of capital in Japan, and the greater diversity in revenue streams compared to American competitors.

With regard to downstream markets, the stimulation of electronic systems industries was an objective of related public policies. To the extent that those measures have provided for an expanded domestic semiconductor market, one can consider them a by-product of industrial promotion, but not *a priori* a function of semiconductor promotion. What emerges from the analysis of Japanese industrial policy for semiconductors is a sense that government policies and private sector strengths were integral to industry development. Factors both related and unrelated to government targeting continue to shape comparative advantage in the industry. Taken together, it is not possible to single out government policy as a make-or-break proposition for the industry. Since structural and cultural factors also played a significant role in determining the competitive position of the semiconductor industry, it is unrealistic to assume that the formula may be translated to an American context.

NOTES

1. Michelle K. Lee, "High Technology Consortia: A Panacea for America's Technological Competitiveness Problems?" *High Technology Law Journal* 6:2 (1992), 353. See also Evelyn Richards, "Chief Sees New Aim for MCC: Consortium Pushed to be Businesslike," *San Jose Mercury News*, 5 May 1991, E1.

2. American antitrust and antimonopoly laws have considered interfirm collaboration as monopolistic and have discouraged the very cooperation and consolidation that is fostered by European governments and Japan. Recognition of this legal restraint encouraged the passage of the National Cooperative Research Act in 1984. This law permits the formation of consortia and cooperative joint research, a development that should promote R&D efficiencies by pooling industry resources.

3. Agencies include the National Science Foundation, Department of Defense, National Security Agency.

4. High-temperature superconductivity (HTS) illustrates the potential problem between commercial and military technology development. The United States has outspent Japan and Europe in its effort to develop this technology, but according to a study by the Office of Technology Assessment, Japan is ahead in the race for commercial application. A large reason is that the government has channeled 70% of its money through DARPA, emphasizing military applications. The Department of Energy has received much of the remaining funds but does not have much experience with working with industry in such matters. See George Lodge, *Perestroika for America: Restructuring U.S. Business-Government Relations for Competitiveness in the World Economy* (Boston: Harvard Business School Press, 1990), 98.

5. Motorola and Texas Instruments have employed VHSIC technology in their ASIC and DRAM development work. David T. Methe, *Technological Competition in Global Industries: Marketing and Planning Strategies for American Industry* (New York: Quorom Books, 1991), 164.

6. Original members included Advanced Micro Devices, AT&T, Digital Equipment Corporation, Harris Corporation, Hewlett-Packard, Intel, IBM, LSI logic Corporation, Micron Technology, Mororola, National Semiconductor Corporation, NCR Corporation, Rockwell International, and Texas Instruments. LSI logic dropped out of the consortium in 1993.

7. U.S. National Advisory Committee on Semiconductors, *Preserving the Vital Base: America's Materials and Equipment Industry* (Washington, D.C.: Government Printing Office, July 1990).

8. U.S. Defense Science Board to the Department of Defense, *Report of the Defense Science Board Task Force on Defense Semiconductor Dependency* (Washington, D.C.: Office of the Under Secretary of Defense for Acquisition, February, 1987), 26-28.

9. Robert White, "Technology and the Bush Administration: Moving Beyond the Industrial Policy Debate," Speech to the National Press Club, Washington D.C., September 29, 1992.

10. U.S. National Advisory Committee on Semiconductors, *A Strategic Industry at Risk* (Washington, D.C.: Government Printing Office, 1989), 18.

11. WSTS statistics, cited in Thomas R. Howell, Brent L. Bartlett, and Warren Davis, *Creating Advantage: Semiconductors and Government Industrial Policy in the 1990's* (San Francisco: Semiconductor Industry Association, 1992), 64.

12. Integrated circuit engineering, cited in Standard & Poor's, "Electronics," *Standard*

& Poor's Industry Survey, June 1994.

13. Report of a Federal Interagency Staff Working Group on the Semiconductor Industry, (Washington D.C., November 1987), cited in Laura D'Andrea Tyson, *Who's Bashing Whom: Trade Conflict in High Technology Industries* (Washington, D.C.: Institute for International Economics, 1992), 106.

14. Infant-industry protection was implicit because American firms had large market shares at home and in third markets but not in Japan. The study suggested that without that protection there would not have been any Japanese producers of 16K chips (1978-1984), the chips could have been more cheaply purchased from American producers, and that the existence of implicit protection resulted in a less than optimal allocation of economic resources. However, the study concluded that the United States lost more than Japan as a result of the protection. See Paul Krugman, "Market Access and International Competition: A Simulation Study of 16K Random Access Memory," in *Empirical Methods for International Trade*, Robert Feenstra, ed. (Cambridge: MIT University Press, 1985).

15. Informal working groups from the two governments had previously discussed ways to alleviate tension and made recommendations accordingly. In 1982 the U.S.-Japan Work Group on High Technology Industries agreed to make American access to the Japanese market comparable to that enjoyed by the Japanese in the American market. The agreement had little teeth and had little to no impact as American market share continued to fall in the Japanese market.

16. Memorandum from President Reagan to U.S. Trade Representative Clayton Yeutter, 51 Fed. Reg. 27,811 (1986), Office of the Press Secretary, White House Fact Sheet, U.S.-Japan Semiconductor Trade Agreement, July 31, 1986.

17. Dumping had ceased in the American market, but the STA covered third markets as well and dumping continued, in this case in Asia. Reagan imposed 100% duty sanctions on $300 million in Japanese imports.

18. Semiconductor Industry Association, *Four Years of Experience Under the U.S-Japan Semiconductor Agreement*, Fourth Annual Report to the President (Cupertino: Semiconductor Industry Association, November 1990); Thomas R. Howell, Brent L. Bartlett, and Warren Davis, *Creating Advantage: Semiconductors and Government Industrial Policy in the 1990's* (San Francisco: Semiconductor Industry Association, 1992), 84-85.

19. The two associations agreed on five desirable actions that included promoting long-term relationships between producers and users; increasing design of foreign products; increasing Japanese purchases of existing foreign products; broadening the base of users and suppliers engaged in the market access effort; and accelerating the startup of foreign participation in the consumer and automotive sectors.

20. Quoted in Thomas R. Howell, Brent L. Bartlett, and Warren Davis, *Creating Advantage: Semiconductors and Government Industrial Policy in the 1990's* (San Francisco: Semiconductor Industry Association, 1992), 89.

21. The agreement abolished the fair market value system created under the original 1986 agreement to prevent dumping. Although the new agreement contained provisions for monitoring semiconductor pricing, there was no trip wire, such as reflecting production costs for enforcement of antidumping rules. Language concerning the 20% figure was more explicit in that it represented neither a floor, ceiling, nor a guarantee. The terms of the agreement would be satisfied if foreign and not just U.S. market share reached 20% of Japanese consumption. Noting the market opening progress, the Bush

Administration removed the $300 million in retaliatory tariffs that Reagan had slapped on Japan. See "A Chip War With Japan? *Business Week*, 22 June 1992, 34.

22. U.S. Department of Commerce and International Trade Administration, *Joint Report of the U.S.-Japan Working Group on the Structural Impediments Initiative* (Washington, D.C.: Government Printing Office, 1990), 43.

23. For comprehensive work on promotion in consumer electronics see Developing World Industry and Technology, Inc., *Sources of Japan's International Competitiveness in the Consumer Electronics Industry: An Examination of Selected Issues*, report submitted to U.S. Office of Technology Assessment (Washington, D.C.: Government Printing Office, June 30, 1980). For more on telecommunications and computers see U.S. Office of Technology Assessment, *Competing Economies: America, Europe, and the Pacific Rim* (Washington, D.C.: Government Printing Office, 1991); U.S. International Trade Commission, *Foreign Industrial Targeting and Its Effects on U.S. Industries, Phase I: Japan*, USITC Publication no. 1437 (Washington, D.C.: Government Printing Office, 1983).

24. U.S. International Trade Commission, Global Competitiveness of U.S. *Advanced Technology Manufacturing Industries: Semiconductors, Equipment and Manufactures*, USITC Publication no. 2434 (Washington, D.C.: International Trade Commission, 1991), 12.

25. Ibid., 17.

26. The U.S. Federal Trade Commission has characterized such episodes as healthy for competition. U.S. Federal Trade Commission, *64K DRAM Components from Japan, No. 731-TA-270* (Washington, D.C.: International Trade Commission, 1985), 72. See also U.S. Federal Trade Commission, *Dynamic Random Access Memory Semiconductors from Japan, No. 731-TA-300* (Washington, D.C.: International Trade Commission, 1986), 6.

27. Japanese companies lost almost $4 billion in 1985, American companies over $1 billion.

28. The STA linked the definition of dumping to production costs, while the GATT defined dumping as selling for less than fair value as indicated by sales in the home market. The STA provision prevents Japanese firms from artificial pricing regardless of home-market values.

29. U.S. Office of Technology Assessment, *Semiconductor Technology Competitiveness* (Washington, D.C.: Government Printing Office, October 1990), 18.

30. U.S. National Advisory Committee on Semiconductors, *A Strategic Industry at Risk* (Washington, D.C.: Government Printing Office, 1989), 10.

31. Japan technology surveys by engineers and scientists have arrived at this general conclusion. Similar assessments have been made by defense-related organizations and other government departments and agencies.

32. Martin Fransman, *The Market and Beyond: Cooperation and Competition in Information Technology Development in the Japanese System* (Cambridge: Cambridge University Press, 1990), 165.

33. U.S. National Advisory Committee on Semiconductors, *A Strategic Industry at Risk* (Washington, D.C.: Government Printing Office, 1989), 19.

34. Gregory Tassey, *Technology Infrastructure and Competitive Position* (Norwell: Kluwer Academic Publishers, 1992), 134. See also Thomas R. Howell, Brent L. Bartlett, and Warren Davis, *Creating Advantage: Semiconductors and Government Industrial Policy in the 1990's* (San Francisco: Semiconductor Industry Association, 1992).

35. U.S. Defense Science Board to the Department of Defense, *Report of the Defense*

Science Board Task Force on Defense Semiconductor Dependency (Washington, D.C.: Office of the Under Secretary of Defense for Acquisition, February 1987). See also U.S. Technology Administration, U.S. Department of Commerce, *Emerging Technologies: A Survey of Technical and Economic Opportunities* (Washington, D.C.: Government Printing Office, Spring 1990).

36. U.S. International Trade Commission, *U.S. Global Competitiveness: Optical Fibers, Technology and Equipment*, USITC Publication no. 2054 (Washington, D.C.: International Trade Commission, January 1988), 8-1 to 8-9.

37. Thomas R. Howell, Brent L. Bartlett, and Warren Davis, *Creating Advantage: Semiconductors and Government Industrial Policy in the 1990's* (San Francisco: Semiconductor Industry Association, 1992), 143.

38. For example, in 1975 American firms accounted for 80% of Japanese market. The VLSI project strengthened the Japanese infrastructure in production and that share fell to 50% by 1980. By 1990 the American share of the Japanese market was only 17%. Michael Borrus, *Competing for Control: America's Stake in Microelectronics* (Cambridge: Ballinger, 1988), 128.

39. VLSI Inc., cited in National Advisory Committee on Semiconductors, *Preserving the Vital Base: America's Materials and Equipment Industry* (Washington, D.C.: Government Printing Office, July 1990), 19.

40. In electroceramics and ceramic packages the Japanese have over 95% of world market share. Ibid., 22.

41. U.S. General Accounting Office, *Federal Research: SEMATECH's Efforts to Strengthen the U.S. Semiconductor Industry* (Washington, D.C.: Government Printing Office, 1991), 2.

42. The withdrawal of American DRAM production during the 1980's hit the SM&E industry particularly hard because DRAM production accounted for nearly 39% of all SM&E sales in the United States. U.S. International Trade Commission, *Global Competitiveness of U.S. Advanced-Technology Manufacturing Industries: Semiconductor Manufacturing and Testing Equipment* USITC Publication no. 2434 (Washington, D.C.: International Trade Commission, September 1991), 24-28. The only major American SM&E survivors have done so by establishing close customer relationships in Asia. U.S. National Advisory Committee on Semiconductors, *A Strategic Industry at Risk* (Washington, D.C.: Government Printing Office, 1989), 12.

43. U.S. National Advisory Committee on Semiconductors, *Preserving the Vital Base: America's Materials and Equipment Industry* (Washington, D.C.: Government Printing Office, July 1990), 7.

44. *Semiconductor materials*: In 1980 Japanese firms held 21% of world market share in semiconductor materials; by 1990 that share had increased to over 73% of world market share. During that same period, American world market share declined from 42% to less that 13%, most of which was sold in the American market. *Optical lithography*: American world market share went from over 70% to under 2% by 1990. When one of the last remaining American producers of lithography equipment was the target of a Nikon takeover bid, a coalition of U.S. firms (including IBM and Du Pont) bought a majority interest in its manufacturing operations. Thomas R. Howell, Brent L. Bartlett, and Warren Davis, *Creating Advantage: Semiconductors and Government Industrial Policy in the 1990's* (San Francisco: Semiconductor Industry Association, 1992), 220. *Automated testing equipment*: American firms went from a dominant position to under 20% while Japanese producers moved into a dominant world market position. For market

figures see U.S. National Advisory Committee on Semiconductors, *Preserving the Vital Base: America's Materials and Equipment Industry* (Washington, D.C.: Government Printing Office, July 1990).

45. Jon Sigurdson, *Industry and State Partnership in Japan: The Very Large Scale Integration Project* (Lund: Research Policy Institute, 1986), 86-89.

46. Ibid., 102.

47. The market position of American materials producers was further eroded owing to continued acquisition of materials companies by foreign buyers. See U.S. National Advisory Committee on Semiconductors, *Preserving the Vital Base: America's Materials and Equipment Industry* (Washington, D.C.: Government Printing Office, 1990), 12.

48. U.S. Defense Science Board to the Department of Defense, *Foreign Ownership and Control of U.S. Industry* (Washington, D.C.: Government Printing Office, 1991), 44.

49. U.S. Congressional Budget Office, *Using R&D Consortium for Commercial Innovation: SEMATECH, X-Ray Lithography, and High Resolution Systems* (Washington, D.C.: Government Printing Office, July 1990), 31.

50. U.S. Congressional Budget Office, *Using R&D Consortium for Commercial Innovation: SEMATECH, X-Ray Lithography, and High Resolution Systems* (Washington, D.C.: Government Printing Office, July 1990), 26-27, 3.

51. U.S. General Accounting Office, *Federal Research: SEMATECH's Efforts to Strengthen the U.S. Semiconductor Industry* (Washington, D.C.: Government Printing Office, 1991), 23.

52. SEMATECH, Annual Report, 1991.

53. Keith Erickson and Ashok Kanagal, "Partnering for Total Quality," *Quality* (September 1992), 17-20.

54. U.S. Congressional Budge Office, *Using R&D Consortium for Commercial Innovation: SEMATECH, X-Ray Lithography, and High Resolution Systems* (Washington, D.C.: Government Printing Office, July 1990), 32-34.

55. U.S. National Advisory Committee on Semiconductors, *Preserving the Vital Base: America's Materials and Equipment Industry* (Washington, D.C.: Government Printing Office, 1990), 4, 16.

56. The Semiconductor Equipment and Materials Institute reported that semiconductor equipment orders increased 23.9% in August 1992 over 1991 levels. See Eric Savitz, "Time for a Change: Semiconductor Equipment Makers Poised for a Turnaround," *Barrons,* 16 November 1992, 35.

57. Standard & Poor's, "Electronics," *Standard & Poor's Industry Surveys*, June 1994, 41.

58. Ibid.; VLSI Research Inc., "The Top Ten Semiconductor Equipment Suppliers for 1991," *Electronic Business,* December 1991, 27. Similar sentiments are expressed by the CEO of Cirrus Logic when he stated that "without [sematech] we wouldn't have a domestic semiconductor equipment industry today. We would be dependent on Japanese equipment." Robert Ristelheuber, "Setting Sun?" *Electronic Business,* April 1994, 58.

59. Part of this is related to the difficulty of raising sufficient long-term capital for R&D and many American suppliers have, therefore, turned to Japanese companies for investment capital. U.S. General Accounting Office, *Federal Research: SEMATECH's Efforts to Strengthen the U.S. Semiconductor Industry* (Washington, D.C.: Government Printing Office, 1991), 23.

60. Ibid., 3.

61. U.S. National Advisory Committee on Semiconductors, *A Strategic Industry at*

Risk (Washington, D.C.: Government Printing Office, 1989), 22.

62. The American semiconductor industry invests 12% of sales revenues in R&D, plant, and facilities; the average for U.S. industry is 4%. See "R&D Scoreboard," *Business Week*, 15 June 1990, 86.

63. U.S. National Advisory Committee on Semiconductors, *Capital Investment in Semiconductors: The Lifeblood of the U.S. Semiconductor Industry* (Washington, D.C.: Government Printing Office, 1990), 4.

64. One Japanese executive commented that "before the bubble economy collapsed we could just issue commercial paper, but now we have to borrow from a bank." See Robert Ristelheuber, "Setting Sun?" *Electronic Business*, April 1994, 58.

65. American corporations rely much more on bonds and other debt securities in debt financing. Debt securities are typically purchased by an anonymous and dispersed group of investors and there is no mechanism to enable such investors to provide additional financial assistance in time of intense competition and declining returns. Moreover, declining credit worthiness means the firm must pay higher yields, thereby increasing its cost of capital at a critical period.

66. U.S. General Accounting Office, *Federal Research: SEMATECH's Efforts to Strengthen the U.S. Semiconductor Industry* (Washington, D.C.: Government Printing Office, 1991), 22.

67. Japan uses the Japan Development Bank and the Japan Key Technology Center. Europe has the European Investment Bank, an EC institution which uses its clout to borrow money on the capital markets and make loans to spearhead technologies at preferred rates.

68. U.S. National Advisory Committee on Semiconductors, *Capital Investment in Semiconductors: The Lifeblood of the U.S. Semiconductor Industry* (Washington, D.C.: Government Printing Office, 1990), 18.

69. MITI funds satellite organizations that take equity ownership in firms pursuing promising technologies. For example, NEDO assumes a 50% equity position in research corporations that it establishes. The Basic Research Center, funded by MITI and the Ministry of Posts and Telecommunications, performs a similar mission. In Europe, national governments (particularly France) have routinely infused large sums of equity capital in large concerns (SAGS-Thomson in France for example). The Flanders Regional Investment Society is a promotional organization in Belgium that facilitates the development of corporations in promising technological areas by providing for its initial capitalization.

70. For more on the role of the capital gains tax and semiconductor investment see David T. Methe, *Technological Competition in Global Industries: Marketing and Planning Strategies for American Industry* (New York: Quorom Books, 1991), 179-182.

71. It takes an American firm four years to accumulate the depreciation tax benefits that Japanese semiconductor firms acquire in one year. See William F. Finan and Chris Amundsen, *Analysis of the Relative Economic Benefits of Tax Depreciation Policies for Semiconductor Equipment and Facilities in the United States and Japan* (San Francisco: Technecon Analytic Research Inc., February 1992), 8.

72. Gregory Tassey, *Technology Infrastructure and Competitive Position* (Norwell: Kluwer Academic Publishers, 1992), 118.

73. Michelle K. Lee, "High Technology Consortia: A Panacea for America's Technological Competitiveness Problems?" *High Technology Law Journal* 6:2 (1992), 347.

74. Clyde V. Prestowitz, ed., *Powernomics: Economics and Strategy After the Cold War* (Lanham: Madison Books, 1991); Clyde Prestowitz, *Trading Places: How We Allowed Japan to Take the Lead* (New York: Basic Books, 1988).

75. Dataquest.

76. Robert Ristelhueber, "U.S. Makers Climb in World's Top 15," *Electronic Business*, January 1993, 17.

77. Also, emergence of rival DRAM producers in Korea at the low end of the scale may eventually challenge the Japanese position in that market. "Current Conditions for Japan's Semiconductor Industry," *Japan 21st Century*, January 1993, 43.

78. U.S. Office of Technology Assessment, *Competing Economies: America, Europe, and the Pacific Rim* (Washington, D.C.: Government Printing Office, 1991), 269.

7

Conclusions and Concepts Revisited

The increasing role of economic and technological factors in the calculus of national security lends new substance to the industrial policy debate. The addition of a strategic dimension has imparted greater urgency to the issue in the United States and elsewhere. Foreign industrial policies have always been a source of concern for American firms. Such policies affect American companies by promoting foreign rivals and limiting access to important markets. However, the perceptible attenuation of America's technological foundation and the attendant risks posed to economic power and national security have elevated that concern to the level of a major public policy issue. The overriding issue at the core of this debate is whether a "business as usual" policy is sufficient to guard national technology assets. Or must the United States government become more involved in commercial competition against rival economies less constrained by principles of laissez-faire?

Government and industry in the United States have generally had what may be considered an adversarial relationship. Fundamental to the American perspective is the notion that government intrusion in the free-market economy is not only inefficient but counterproductive. Industrial policy in the United States conjures up images of fat bureaucrats meddling in affairs best left to the market. In principle, government intervention in specific industrial sectors is justified in two respects: on the grounds of protecting an important economic asset in the interest of national security, and in response to unfair trade practices. For the most part, government policy has the objective of creating favorable macroeconomic conditions for industry to convert innovation into commercial profit.

Since the early twentieth century the United States has been the world's preeminent economic superpower, a position reinforced by the economic emasculation of Europe through two successive world wars. The United States held virtually every economic advantage, leading in productivity, technology, natural resources, and manufacturing. As the new economic superpower, the United States became the leading proponent of free trade and laissez-faire economic principles. Such an orientation carries obvious benefits to the power that can produce more goods more cheaply than any other competitor. The subsequent loss of comparative advantage in labor-intensive activities was part of the natural progression of economic activity into higher technological and capital-intensive industries. However, in the past two decades the United States has been challenged in increasingly technology-intensive activities, and more recently the most advanced technology industries have suffered in international competition. Although there is a growing consensus that something must be done to safeguard important economic assets, the United States has been slow to embrace the concept of industrial policy.

The American government does not match the comprehensive support that the Japanese and European governments provide their high technology networks through coordination of various R&D agendas and subsidization of technology development. SEMATECH is the application of such support, but only to a single area in the semiconductor industry. A presumption that the American government has already accepted responsibility for promoting important commercial technologies is, therefore, premature. Even in the American context, these measures as policy precedent are less radical than one would think. The trade agreements made managed trade an operational reality, but the government has always had a legitimate role in responding to unfair trade practices. The SEMATECH initiative was passed under the aegis of a broader defense bill, and the opponents to industrial targeting deferred on the grounds of national security. Both venues represent a long-accepted justification for government intervention. The existence of these two measures does not *a priori* mean that government policy for the promotion of critical high technology sectors has been institutionalized as a matter of economic policy.

It is generally accepted that the military threat confronting the Western world and Japan since 1945 has diminished considerably with the dissolution of the Soviet Union. At the same time, however, as the war in the Persian Gulf aptly demonstrated, military power will remain a critical element in providing for national security. A strong economy has always been vital to defense, and economic sectors that provide important technologies and goods in maintaining military preparedness will, therefore, remain fundamentally important. But the role of the economy is becoming increasingly direct and central in an era given to economic concerns in a single superpower world. Recognizing that the nature of those threats are as much economic as they are military, the Clinton administration raised the principle of economic security to the level of national security status.

Examining the proper role for government in protecting economic assets is a multifaceted task. First, it should be recognized that the American military has functioned as a sub-rosa set of development policies where militarily vital technologies are concerned. As a result, the dismantling of the Cold War military structure will have a large impact on the domestic electronics industry and other sectors that have benefited from the patronage of Pentagon procurement. For example, in 1992, 21% of all American-related electronics manufacturing output was defense related, compared to only 11% in Europe and 2% in Japan.[1] Second, technologies subsidized by government contracts have become increasingly esoteric and have had a decreasing impact on commercial competitiveness. The contraction of military spending will affect the cash flows of certain companies, but ending a reliance on government patronage may free up resources to compete more effectively in commercial markets. Third, the Pentagon has become more dependent on dual-use commercially developed technology. The Defense Science Board concluded that the Pentagon is dependent not only on commercially developed technology but on commercially developed foreign technology. Numerous examples of strategic withholding by Japanese firms indicate that reliance on foreign supplies of strategic components creates an opportunity for foreign manipulation. These realities frame the debate on how government may best respond to new demands on national power. In the face of economic competition, the needs of the military and the economy have converged over the issue of how best to generate commercial innovation. It is not possible to support all high technology industries. The government must select certain industries over others, and favor certain high technology activities by assessing the social return of innovation and its importance in the scheme of national security. Selecting an industry that is economically strategic on the rather vague criteria of abnormal profits and/or positive externalities, while not impossible, is very difficult. Advocates of industrial policy submit that it is better to risk supporting an industry that is falsely judged to be strategic than to allow the collapse of others that are falsely judged not to be. However, promotion carries a price. It taxes government treasuries already stretched to the limit and there is the risk that government intervention can have neutral or even negative consequences for the industry. Industrial policy in Europe illustrates the difficulties in targeting and the potential for billions of dollars in misallocated resources; and as the semiconductor case study illustrates, the utility of government targeting should not be overstated. Moreover, it is impossible to draw general conclusions on the efficacy of government promotion without reference to related factors affecting competition.

LESSONS OF THE SEMICONDUCTOR INDUSTRY

One of the research questions to be explored was the extent to which foreign industrial policies have been successful and how this has affected the American industry. Is the decline in the American semiconductor industry the result of free-market competition or has government shaped the competitive outcome in important markets? That a concerted policy of industrial targeting through trade policies and structural manipulation has occurred in Japan, Europe, and elsewhere is difficult to refute. However, the utility of government policy in international competition should not be overstated. The case of Japan suggests that industrial policy accelerated the development of the industry during the 1970's, but subsequent competitive strength stemmed significantly from macro-economic strengths and demand-pull. In Europe, national government policies actually retarded the development of a competitive semiconductor industry through the 1970's and early 1980's. Although policy instruments such as technology push and import restrictions were similar to those utilized by Japan, the blend was different and the results less productive.

To attribute the Japanese challenge in semiconductors to implementation of industrial policy is therefore an oversimplification. Yet the recognition that indigenous factors were fundamental to the success of that policy does not mean that government policy itself was unimportant. Government targeting was an enabling mechanism. The protection and promotion of the industry during the 1970's quite clearly positioned it to become a world leader. The American industry at that time had the most sophisticated technology and manufacturing processes, and led in world market share. American devices were superior and cheaper. Together with scale and first-mover advantages, there was no indication that the American position would suffer in free-market competition. Without comprehensive protection from investment and imports, the creation of a domestic demand base for Japanese producers and the movement into the higher technological echelon of the industry would have been extremely difficult. In addition, *keiretsu* demand linkages alone would have been an insufficient impetus because the more sophisticated components required in consumer electronics and computers were available only from American firms. Japan generally imported only those advanced components it did not yet produce, and then later substituted domestic production for imports as soon as domestic capacity came on line; in this regard, American companies could retain their market position in Japan only by shifting the mix of goods that Japanese producers could not yet produce. Without government promotion, the Japanese industry would have been denied the captive home base and capital to move into more advanced stages of the industry.

Together, government policy and domestic market structure, and the business practices they fostered, provided substantial advantages to Japanese producers in international competition. Some of the most prominent reasons for Japanese success during the 1980's, such as dumping, the persistence of a closed domestic market, advantages in capital formation, and the strength of downstream industries, are not specifically related to industrial policy. MITI has been more important as a facilitator in the past ten years, supporting and organizing research efforts in next-generation technologies that may later define comparative advantage in the industry.

Since European semiconductor policy has had a regional focus without much attention given to sale on global markets, the best way to determine the affect of industrial policy on the American industry is local market share. Through the

1970's American firms held slightly over half of the European market and that share decreased only slightly through the 1980's to 45%. European promotional measures have not made any substantive impact on the American industry, and most of the market share lost by American firms can be attributed to a larger Japanese position and not to European policies.

America's strong market position in Europe has been an important part of its overall strength in world market share. New directions in European policy and a commitment by member governments to information technology indicate the capacity and the will to sustain a European-owned semiconductor industry. Trans-European R&D projects, tariffs, and changes in rules of origin may ultimately chip away at the market position of American firms by tilting the advantage toward European producers. However, there is no indication that the American semiconductor producers will be seriously challenged by the European industry. Rules of origin changes will most likely affect the Japanese industry more, since the United States already has a higher level of European production. Given the limited results of European technology initiatives, American firms should retain their dominant position over the European industry in Europe, especially if American producers maintain a high level of foreign investment. On the other hand, Japanese competition in Europe has intensified. Although Japanese producers have lost their grip on the European market in memories,[2] it is more likely that the Japanese industry and not European producers will be the ones to mount the broadest challenge to the American position.

Regional and EC efforts, though very recent, offer little hope of ever seriously challenging the major semiconductor producers. An important element in the European policy has been the extension of a European production base. In the absence of a domestic industry capable of meeting European demand, the development of a foreign-owned European production capability may be viewed as a second-best alternative. Nonetheless, local production grants European authorities greater control over strategic components. It makes it more difficult for foreign-based companies to coordinate their actions, and it eases national security concerns. If a company acting as a surrogate for the home nation withholds a vital chip to exert leverage on another's foreign policy, then the local government may, if necessary, simply occupy the facility. An important area of further research would be to examine to what extent a foreign owned local production base is an adequate substitute for a competitive national semiconductor industry. To what extent does foreign ownership retard or facilitate the flow of externalities to the local economy? A related question is the extent to which national security concerns remain.

The role of government promotion continues in a diminished capacity in Japan at the same time promotion appears to be gaining in Europe and the United States. The Japanese case reflects the development of a world-class industry that no longer requires government assistance. European policies reflect a recognition of previous failures and a greater sense of urgency to develop a competitive semiconductor industry. The competitive problem for the American industry stems from a combination of foreign policies and certain disadvantages not attributed to government policy. The semiconductor case suggests that American public policy should recognize that whether industrial policy succeeds or fails depends on the interplay of structural and cultural factors. Beyond that, the lack of competition in the European case and intense competition in Japan indicate the importance of competition to any successful industrial strategy. Protection from the discipline of the market through trade policies, procurement, and or subsidy only serves to camouflage inefficiencies unless it is integrated into a general pro-market strategy. The objective of any policy should be economic efficiency as measured by cost and price competitiveness on world markets.

The steady challenge to America's position in microelectronics is a legitimate policy concern. Yet the strategic value of an industry is not sufficient reason to intervene if that industry were capable of successfully adjusting to new competitive conditions without government assistance. The American industry is clearly not moribund. American firms regained the lead in world market share and the industry shows signs of continued prosperity through 1994, with forecasts of double-digit growth.[3] However, competitive factors supporting a long-term advantage for Japan suggest it is questionable whether the American industry has the ability to accomplish an effective adjustment on its own. The American semiconductor industry has suffered from a combination of limitations imposed by industry structure, macroeconomic constraints, and market access impediments. Any government policy to assist the industry must have components that address these areas. Trade remedies alone are inadequate to the task of ensuring a competitive industry because managed trade and antidumping laws do not address distinctly domestic problems. Government should be prepared to augment private resources through industry-government consortia or tax policy to ensure that important technologies do not go underfunded. Recent efforts by the American government to spur competitiveness and economic growth in the industry have been effective and successful. They have established an important role for the government in enhancing the long-term strength of the industry.

INDUSTRIAL POLICY

Trade Policy

From a public policy perspective the European market remains important, but is a source of less concern compared to the effect that the Japanese trade structure has on American trade. This is not to suggest that the manipulation of the free flow of goods and investment through less transparent policies is necessarily better. However, the strong American position in the European market for semiconductors and an aggregate trade surplus with the EC together reduce the imperative for a public policy initiative in the European theater. By contrast, the level of American (and foreign) market share in Japan remains low, and it is not due simply to the fallout from free-market competition. It is a situation that does not reciprocate the market access Japanese producers enjoy in the United States. Various structural and market-related reasons have been advanced to account for low foreign semiconductor share, and despite the apparent efforts by MITI, it continues to be very difficult to reach the 20% commitment. MITI's inability to guarantee an American share of 20% suggests either its efforts are symbolic and not substantive or MITI may not be capable of such direct influence anymore.

The Japanese trade surplus in semiconductors is not unique since the economy generates enormous trade surpluses with the rest of the world in most manufactured goods. Even if one attributes this surplus to superior organization, it is evident that Japan has managed to avoid complete integration into the system of liberal international trade fashioned after World War II. The persistence of a closed market has a fundamental effect on semiconductor competition. It reserves demand for Japanese companies thus guaranteeing economies of scale to cover investment costs, and *keiretsu* market arrangements provide a hedge against periods of overcapacity while facilitating a rebound from recessions. These factors reduce the risk of investment in production capacity and can lead to excess capacity that is translated into export drives on world markets.

The Japanese system, therefore, not only denies American firms an important source of market share but has a tendency to encourage dumping. The trade regime of the United States has been incapable of dealing with these disruptions caused by the Japanese industry. In principle, the government prosecutes "unfair" trade cases (pursuant to article 301 and super 301), and provides relief for injured parties, but it is not a structure geared toward projecting a broadly conceived response in defense of an industry subjected to a systemic foreign challenge. The 1988 Omnibus Trade and Competitiveness Act strengthened the original 301 clause of the 1974 trade law, making it easier to seek damages and provides for tougher penalties. However, anti dumping laws take too long to implement and since they only reestablish fair market value, they may prevent future damage but cannot repair the affects of past dumping. American trade law operates from a static orientation, one that considers firm behavior at the time of the petition while not fully addressing the systemic causes leading to dumping.[4]

The antidumping provisions of the semiconductor trade agreements recognized the debilitating effects of dumping and compelled Japanese producers to make components reflect production costs. The STA did stop the dumping; applied provisions to third markets in part to prevent the migration of those user industries to cheaper offshore sourcing locations. But higher DRAM prices did not resurrect the American DRAM industry, it merely raised production costs for those downstream industries that depended on those components. The STA dealt with the industry within a general strategy, and it has discouraged dumping since

1986. In this regard, the antidumping provisions have been a success. However, its failure to repair damage inflicted on an important part of a critical industry suggests that the approach to dumping and other forms of unfair trade in high technology is in need of measures that not only discourage dumping with severe punitive measures but also provide a way to restore the industry balance that preceded the dumping.

The persistence of structural impediments and policy initiatives in Japan and Europe provide the justification for an American response not based on free-trade principles. This is not a call for blanket protectionism. Neither is it an indictment of free trade. Several new econometric studies by advocates of the new trade theory support the belief that trade liberalization remains in the national interest even under conditions of imperfect and strategic competition.[5] Moreover, conclusions derived from our analysis on strategic trade theory indicated that it is not sufficiently rigorous to serve as a policy tool. Given the complexities of real-world competition, it is unreasonable to assume that "profit shifting" through government intervention will bring the rewards suggested by recent currents in international trade theory. Accordingly, the United States should not provide export subsidies and should not erect trade barriers to protect the domestic market from international competition. Rather, a primary task for the American government is to open up the maximum foreign markets for American high technology goods. The American government should remain committed to free trade, but must also be prepared to intervene in the form of reciprocal free trade or managed-trade agreements to open up markets constrained by structural and government impediments to open competition. In interviews with American semiconductor executives, this was said to be the most compelling role for government in promoting semiconductors.[6]

A criticism often levied on managed trade is that it is not economically efficient, and it can camouflage competitive deficiencies that would otherwise be corrected. However, the semiconductor agreements do not conceal competitive weakness, rather they have promoted efficiencies for the American industry. As noted earlier, a major advantage for the Japanese industry has been an enormous captive market. This competitive strength for America's principal rival is a competitive weakness for the American industry. Legal terms that break or weaken that monopoly and open up markets for American producers will promote lower production costs and revenues that may be reinvested in production and R&D. Another criticism is that managed trade cartelizes markets and increases prices by limiting competition. Accordingly, it has been suggested that the STA was responsible for the chip shortage of 1987 to 1989 because the antidumping provisions of the STA and the establishment of a price floor on chips encouraged the formation of a Japanese chip cartel.[7] It should be noted, however, that a price floor was not established on chips; semiconductor prices simply had to reflect production costs. The precipitating event was not the STA. Rather, it was the marginalization of non-Japanese DRAM producers. MITI set voluntary production guideposts for DRAMs and EPROMs. Japanese producers conformed to the guideposts in DRAMs, but because non-Japanese competition existed in EPROM's their output did not conform to the MITI guidelines. The market for EPROMs was subject to the same legal provisions under the STA but there was no shortage and prices declined along the learning curve in a predictable fashion, precisely because American firms retained 50% market share. The semiconductor trade agreements have not promoted cartels and limited competition. On the contrary, the market-access provisions have been procompetitive by increasing competition in the Japanese market. Although American market share has not yet reached the agreed level, the agreements have increased that share, and they represent the best vehicle for American companies to overcome impediments in the Japanese market.

Domestic Policy

Competition in the global semiconductor industry highlights some structural weaknesses in the American system of technological development. One such flaw is inadequate strategic direction. Although SEMATECH and the SRC have made efforts toward an industry-wide agenda, SEMATECH has a limited mission in the SM&E markets and the SRC lacks the institutional muscle to implement a comprehensive strategy. Because government participation in collaborative planning is still quite tentative in the United States, there remains no generally accepted mechanism to organize and direct a collective research effort toward a specific goal. Company laboratories, universities, and defense organizations will continue to pursue independent research agendas because this is an important element in competition. At the same time, this can lead to redundancy and to potentially serious gaps in research across a technological front.

Another major weakness in the American development of technology concerns capital formation. The development of advanced microelectronics is a task requiring enormous commitments of capital, equipment, and expertise. It is true that most enterprises at one time or another experience problems of insufficient financing. Capital investment flows are directed in venues that maximize private return with little concern for any inherent strategic value. The result is that risky technologies tend to be underfunded. While European and Japanese firms operate in an institutional framework that lessens the problems of capital formation, the United States has not developed an institutional apparatus to augment private investment in specific R&D. Moreover, in an era of constrained resources, lack of strategic direction makes problems of capital shortfall more acute.

Capital investment has been described as the lifeblood of the semiconductor industry. The advantages held by Japanese producers in capital formation are enhanced by vertical integration and diversified revenue streams. This advantage enables Japanese producers to maintain investment in technology and manufacturing in periods of slack semiconductor demand. Escalating development costs in technology generation and production facilities are a particularly heavy burden for American merchant firms, which do not have the same diversified revenue streams or capacity to develop capital reserves. Although the merchants have a record of being more innovative in response to changing market demands, they cannot match Japanese manufacturing advantages and staying power. Whatever dynamic advantage that American merchants have in advanced chip design is jeopardized by capital requirements that are more easily met by larger electronic producers. This may threaten America's leading position in microprocessors, which are entering an era of commodity pricing.[8] Since investment in R&D and new manufacturing systems is integral to the success of American semiconductor producers, the question is how to make the capital available for requisite investments.

The promotion of precompetitive technologies is an element of government intervention that is most practical for the United States. Such policies are aimed at removing bottlenecks in the development of the generic technologies that underlie commercial development and production. This is not to suggest that the centralization of the R&D agenda would devolve exclusively to government agencies. In addition, it is not a matter of throwing money at the problem. After all, the level of public funding in Japan is a fraction of what it is in Europe and while Japanese producers are at the technological frontier, European producers struggle to catchup. Both the Japanese and European cases illustrate, in different ways, that government and industry must work together to identify the essential elements affecting the technological competitiveness of the industry. An essential ingredient is to maintain a promarket approach so that competitive behavior will translate generic advances into economic efficiencies on the world market. Targeting precompetitive technologies should be part of a larger institutional mechanism to coordinate and clarify promising avenues of R&D. Promotion at this level would augment capital resources, rationalize the R&D agenda, and help meet the race for next-generation technologies that may shape future competitive advantage.

One fear of subsidies in any industry is that they insulate firms from the discipline of the market and breed an operating dependence on that subsidy, making it difficult for government to revoke the subsidy at a later date. Subsidies in the European industry have done little to suggest that such skeptics are wrong. Subsidies are not a substitute for sound management practices. However, several distinctions must be emphasized. First, subsidizing selected technology development is not a production subsidy. A government subsidy matching a private investment in R&D enhances the ability of firms to compete more effectively because it facilitates innovation by organizing resources to build upon the technology infrastructure. It is generally agreed that society under-invests in R&D, and this is most evident in precommercial technologies in which the link between investment and firm appropriation is most uncertain. Second, the subsidy need not necessarily be anticompetitive. An important element in the Japanese success has been the competitive race between participating firms to develop the jointly researched precommercial technology.

Third, the absolute amount of the subsidy is potentially less important than the rationalization it encourages in the industry. When SEMATECH was first proposed in 1986, a dissenting voice in the Office of Management and Budget stated that "the total R&D budget for the 14 semiconductor firms is over $4.5 billion and they claim SEMATECH is critical to their success; yet unless government puts in $100 million a year, it won't fly. Where are their priorities?"[9] The commitment by government to promote semiconductor equipment technology was an organizing force that tapped company-level resources that would not have been committed in the absence of government intervention. The industry recognized a critical technology gap, but from an individual firm perspective, the risk of investment in technology development outweighed its presumed return to the firm. Subsidization of important technologies may or may not be critical for the future health of an industry, but it need not necessarily promote anticompetitive behavior or an inefficient industry structure.[10]

The apparent success of the semiconductor trade agreements and SEMATECH, achieved with a relatively modest investment, lends greater integrity to similar efforts in other areas of strategic importance. These two policies indicate that industrial targeting can be practiced in the United States. SEMATECH embodies several features that reduce the likelihood of political and economic boondoggle. First, SEMATECH is market driven and industry led. There is no presumption that government will dictate the direction of resources; rather, the government facilitates the flow of resources into areas deemed important in conjunction with a cross section of industry sources. Second, 50/50 cost sharing places private funds at risk and imposes a discipline that total subsidization might not achieve. Third, the consortium has rationalized the SM&E industry through the R&D agenda, uniform standards, and improved communication between chip and equipment producers. This major contribution in strategic direction cannot be quantified in monetary terms. In this way, the government provided the framework for industry cooperation, and with partial support, facilitated the process of technological development.

It is important to consider that industrial policy is not a panacea for a more general lack of competitiveness. Industrial policy can influence the composition of factors of supply and factors of demand, but American high technology firms depend upon a healthy macroeconomic environment. Since government cannot control all those factors affecting competitive position in the semiconductor industry, it is clear that there are limits to what one might expect from government promotion. For example, the ability of the Japanese industry to sustain a strong position is not directly attributable to government policy. In Europe, government policy has proved to be very expensive and it has not yet provided any discernible advantage for European producers. At the same time, as the Japanese and American cases suggest, government may have a relatively small but influential role to play. It is perhaps coincidental that there are no other examples of American industries that have recovered dominant world market position once lost to Japanese producers. Given the intensity of foreign competition there was no indication that American SM&E firms were capable of matching Japanese technology and markets; that the Japanese trend toward greater global market share would abate; or that American semiconductor companies would increase their share of the Japanese market. Although market share and technological innovation will be fundamentally determined by the competition and interplay among private sector participants, industrial policy will continue to be a factor in international competition and has been shown to work in Japan, and even the United States.[11]

CONCLUDING THOUGHTS

The purpose of this study was to examine high technology industrial policy within the framework of the semiconductor industry and its relation to economic competition and national security in the post-Cold War world. The semiconductor industry symbolizes the conflict associated with government-sponsored advantage, and lessons from this industry may be more generally applied to other high technology industries that share the three fundamental needs of guaranteed markets, high levels of R&D, and abundant capital. It is reasonable to conclude that government intervention in a strategic industry is justified on grounds of national security if that industry is incapable of making adjustments on its own as a result of foreign industrial policies or constraints imposed by the domestic economy. Government intervention should be limited. It should not replace the discipline of the market with subsidy. It should conjoin private funds and expertise with limited public resources to encourage industry rationalization, to promote precommercial technology development, and to preserve and extend unfettered free trade. The cost of promotion in pre-competitive technologies could be partly offset through a more judicious organ-ization of existing government promotion policies reviewed in Chapter 3.

The nature and extent of government promotion in microelectronics and other high technology industries should be the subject of public debate. The underlying principles of that debate should recognize the strong linkages that high technology bestows on the economy and their impact on general economic health through productivity, wages, skill formation, and investment in R&D; that a successful industrial policy is context dependent; and that measurements of success must also consider the motivating context of promotion. Europe has pumped enormous funds into microelectronics without commensurate returns in technology and market share, but policy makers regard the cost of maintenance as less than the cost of losing the industry altogether. Control over critical technological components carries obvious implications for national security. At the same time, it should be recognized that there is no broad agreement over what is and is not a critical technology. For example, the Critical Technologies Panel has identified 22 technologies critical to economic and military security, but the Council on Competitiveness has identified 94 such technologies. There is a need to establish specific guidelines to create an accepted standard, and there is a need to develop some mechanism by which dangerous reliance may be calculated and acted upon.[12]

The foregoing does not mean that widespread promotion is desirable or problem free. Industrial policy must be used with extreme caution and selectivity. Dozens of industries have already begun lobbying efforts to elicit government protection based on their contribution to national security.[13] Political favoritism and special-interest groups will always have the potential to influence any government program. Accordingly, industrial policy should promote only specific technological goods—ones that promote strong linkages with the rest of the economy and are regarded as critical for military and economic security. The case for such promotion is particularly warranted when access to the product or technology can be manipulated by a few foreign suppliers. By extension, a major task is to determine where the combination of market failure, external benefits, and national interest together provide reasonable justification for government intervention. R&D subsidies should be considered on a case-by-case basis and the creation of a nonpartisan, independent evaluation panel would reduce the possibility of private and political abuse.

From a military perspective, the United States has three alternatives: (1) subsidize a defense-industrial base to satisfy defense requirements; (2) ensure that the industry supplying technology employed by defense remains commercially competitive; and (3) source military components overseas. The first alternative is prohibitively costly. It is a policy open to abuse and low-volume production, and one in which mission-specific standards will make final goods exceedingly expensive. The problems associated with the third option have been examined. The most viable alternative is to ensure that vital defense-related industries are competitive and profitable in world competition. This would provide economies of scale on the commercial side that could, in some measure, reduce per unit costs for the military; ensure that the vital COTS technologies are domestically produced; and guarantee that the United States would have every expectation to supply the technologies necessary for military security.

The traditional role adopted by the federal government toward commercial innovation has been through funding basic research and creating the macroeconomic conditions conducive to business development and economic growth. A more forceful government role in the innovation process had been considered unnecessary since American technological leadership has remained undisputed. However, foreign competition has placed important segments of the American high technology establishment in jeopardy. There are signs that the government-industry relationship is becoming less antagonistic and the way they interact in the development of new technology is changing.[14] The process to safeguard technological assets has already started with SEMATECH and the semiconductor trade agreements, and more generally with the Advanced

Technology Program, the CRADAs, and passage of the National Cooperative Research Act. In the 1990's the United States finds itself locked in an intense global economic competition that has profound implications for the security structure of the country, and the ability to provide for that security will be enhanced through closer coordination of government and private industry.

NOTES

1. "Can the Electronics Industry Survive the Cold War?" *Electronic Business*, January 1993, 68.

2. The Korean producer Samsung Semiconductor has garnered 25% of the memory market at the expense of Japanese producers that once had 80% of that market. By contrast, Siemens has only a 15% share. Andrew Pollack, "A Chip Powerhouse is Challenged," *New York Times*, 17 December 1992, D1.

3. Robert Ristelhueber, "Another Pleasant Surprise This Year for Chip Makers" *Electronic Business*, January 1993, 95-96.

4. This may be attributed to the fact that, while American economic policy considers protection and dumping unfair trade practices, in the long term, they have been perceived to be self-defeating commercial tactics. Dumping has been considered a boon to consumers. Cartelization by definition is a grouping that restricts output and raises prices, and this cannot be viewed as detrimental to American competitive interests, which may underprice such a grouping. Protection has traditionally been associated with inefficiency, shielding uncompetitive enterprises from the rigors of market competition. In the past, the American trade regime has recognized that such practices are annoying but in the end do not appreciably affect the competitive position of American firms.

5. David Richardson, "New Trade Theory and Policy a Decade Old: Assessment in a Pacific Context," Paper presented at the First Australian Fulbright Symposium on *Managing International Relations in the 1980's*, Canberra, March 1992; cited in Laura D'Andrea Tyson, *Who's Bashing Whom? Trade Conflict in High Technology Industries* (Washington, D.C.: Institute for International Economics, 1992), 255.

6. William Gsand, executive vice president of Hitachi America, personal interview, 12 August, 1992. Richard Sanquini, senior vice president of business development at National Semiconductor, personal interview, 11 August 1992.

7. "Economics of Managed Trade," *The Economist*, 22 September 1990, 17.

8. Software is increasingly being run independently of any particular microprocessor. This will allow customers to pick any microprocessor as long as it has sufficient horse power to run the program. As microprocessors enter an era of standard commodity pricing, a major challenge will be posed to those firms that have relied on proprietary architectures to maintain market share.

9. Thomas R. Howell, Brent L. Bartlett, and Warren Davis, *Creating Advantage: Semiconductors and Government Industrial Policy in the 1990's* (San Francisco: Semiconductor Industry Association, 1992), 93.

10. Andy Grove, the CEO of Intel, credits SEMATECH with bringing the industry back from the brink of the latter 1980's. See "Inside Intel," *Business Week*, 1 June 1992, 86.

11. A spin-off of the SEMATECH consortium has been the development of new lithography technology that challenges the global monopoly of Japanese suppliers. IBM's

access to the new technology (in conjunction with the buyout of the Perkins Elmer lithography business) gave it leverage in the joint venture with Siemens and Toshiba to manufacture 16M DRAMS in the United States. America will benefit from research monies and jobs for scientists and engineers. Andrew Pollack, "I.B.M.-Toshiba Joint Chip Venture," *New York Times*, 22 June 1992, D3. See also John Markoff, "Unable to Beat Them, I.B.M. Now Joins Them," *New York Times*, 6 July 1992, D1.

12. Moran suggests that a national security threat exists whenever four countries or four firms supply more than 50% of the world market for a product deemed essential for national defense. Theodore H. Moran, "The Globalization of America's Defense Industries: Managing the Threat of Foreign Dependence," *International Security* 15 (Summer 1990), 57-100.

13. Keith Bradsher, "Industries Seek Protection as Vital to U.S. Security," *New York Times*, 19 January 1993, D1.

14. For example, Washington and Detroit automakers have pooled resources to devise new automobile technology. See Matthew Wald, "Government Dream Car," *New York Times*, 1 October 1993, A1.

BIBLIOGRAPHY

Abbot, T. A. "Measuring High Technology Trade." *Journal of Economic and Social Measurement* 17 (May 1991): 17-44.

Abernathy, William, and Robert Hayes. "Managing Our Way to Economic Decline." *Harvard Business Review* (July/August 1980): 67-84.

Abramovitz, Moses. "Rapid Growth Potential and Its Realization: The Experience of Capitalist Economies in the Postwar Period." In *Economic Growth and Resources*, Edmond Malinvaud, ed. London: Macmillan Press, 1979.

Adams, William J., and Christian Stoffaes, eds. *French Industrial Policy*. Washington, D.C.: The Brookings Institution, 1986.

"America Wants Japanese Science." *The Economist*. 30 January 1988: 68.

"American Technology Policy." *The Economist*. 25 July 1992: 21-23.

Arnold, Walter. "Science and Technology Development in Taiwan and South Korea." *Asian Survey* (April 1988): 34-41.

Aschauer, David A. *Public Investment and Private Sector Growth* (Washington, D.C.: Economic Policy Institute, 1990).

———. "Infrastructure: America's Third Deficit." *Challenge* (March/April 1991): 22-27.

Audretch, David B., Leo Sleuwaegen, and Hideki Yamawaki, eds. *The Convergence of International and Domestic Markets*. Amsterdam: Elsevier Science Publishers, 1989.

Ayres, Robert U. *The Next Industrial Revolution: Reviving Industry through Innovation*. Cambridge: Ballinger Publishing Company, 1984.

Baatz, E. B. "Chips Lead U.S. Recovery." *Electronic Business*. January 1993: 7.

Badaracco, Joseph L., and David B. Yoffie. "Industrial Policy: It Can't Happen Here." *Harvard Business Review* (November/December 1983): 88-99.

Bailey, Martin, and Alok Chakrabarti. *Innovation and the Productivity Crisis*. Washington, D.C.: The Brookings Institution, 1979.

Bailey, Martin, Robert Gordon, William Nordhaus, and David Romer. "The Productivity Slowdown, Measurement Issues, and the Explosion of Computer Power." *Brookings Papers on Economic Activity* no. 2 (1988): 347-431.

Baldwin, Richard, and Paul Krugman. "Industrial Policy and International Competition."

In *Trade and Policy Issues and Empirical Analysis*, Richard Baldwin, ed. Chicago: University of Chicago Press, 1988.

Balk, Alfred. *The Myth of American Eclipse: The New Global Age.* New Brunswick: Transaction Publishers, 1990.

Ballard, Charles L., John B. Shoven, and John Whalley. "General Equilibrium Computations of the Marginal Welfare Costs of Taxes in the United States." *American Economic Review* 75 (1988): 128-138.

Ballard, Steven. *Innovation Through Technical and Scientific Information: Government and Industry Cooperation.* New York: Quorom Books, 1989.

Bartlett, Bruce. "America's New Ideology: 'Industrial Policy'." *The American Journal of Economics and Sociology* 44 (January 1985): 1-7.

Beije, P. R., J. Groenewegen, I. Kostoulas, J. Paelinck, and C. Van Paridon. *A Competitive Future for Europe? Towards a New European Industrial Policy.* London: Croom Helm, 1987.

Bell, Trudy E. "Japan Reaches Beyond Silicon." *IEEE Spectrum* 22 (October 1985): 42-54.

Beltz, Cynthia A. *High-Tech Maneuvers: Industrial Policy Lessons of HDTV.* Washington, D.C.: The American Enterprise Institute, 1991.

Berkman, Barbara. "Start-Up No More, Jessi Gets Down to Business." *Electronic Business.* 18 May 1992: 68-72.

Bernheim, Douglas, and John Shoven. *Comparison of the Cost of Capital in the United States and Japan: The Role of Risk and Taxes.* Stanford: Stanford University Press, 1989.

Bissell, Richard E., and Gordon H. McCormick, eds. *Strategic Dimensions of Economic Behavior.* New York: Praeger Special Studies, 1984.

Blank, Stephen, and Paul Sacks. "If At First You Don't Succeed, Don't Try Again: Industrial Policy in Britain." In *Industrial Vitalization:Toward a National Industrial Policy*, Margaret E. Dewar, ed. New York: Pergamon, 1982.

Blau, John R. "Europe Stumbling in Semiconductor Race." *Research and Technology Management* (March/April 1992): 3-4.

Borrus, Michael. *International Competition in Advanced Industrial Sectors: Trade and Development in the Semiconductor Industry.* A Study prepared for the use of the Joint Economic Committee, Congress of the United States. Washington, D.C.: Government Printing Office, 1982.

———. *Competing for Control: America's Stake in Microelectronics.* Cambridge: Ballinger, 1988.

Borrus, Michael, James Millstein, and John Zysman. *U.S.-Japanese Competition in the Semiconductor Industry: A Study in International Trade and Technological Development.* Berkeley: Institute of International Studies, 1982.

Bosworth, Barry P., and Robert Z. Lawrence. "America in the World Economy." *Brookings Review* 7 (Winter 1988/1989): 39-48.

Bradsher, Keith. "U.S. Escalates War of Words Over Chip Imports From Japan." *New York Times.* 30 December 1992: D1.

———. "Industries Seek Protection as Vital to U.S. Security." *New York Times.* 19 January 1993: D1.

———. "High-Tech Industry is Hard to Help With Subsidy." *New York Times.* 2 February 1993: C1.

Brander, James A. "Rationales for Strategic Trade and Industrial Policy." In *Strategic*

Trade Policy and the New International Economics. Paul Krugman, ed. Cambridge: MIT Press, 1986.

———. "Shaping Comparative Advantage: Trade Policy, Industrial Policy, and Economic Performance." In *Shaping Comparative Advantage*, R.G. Lipsey, and W. Dobson, eds. Toronto: C.D. Howe Institute, 1987.

Brander, James A., and Barbara Spencer. "Export Subsidies and International Market Share Rivalry." *Journal of International Economics* 18 (February 1985): 83-100.

Branson, William H. "Industrial Policy and U.S. International Trade." In *Toward a New U.S. Industrial Policy?* M. L. Wachter. ed. Philadelphia: University of Pennsylvania Press, 1983.

Broad, William. "Defining the New Plowshares Those Old Swords Will Make." *New York Times.* 5 February 1992: A1.

Burton, Daniel F., and Victor Gotbaum, eds. *Vision for the 1990's: U.S. Strategy and the Global Economy.* Cambridge: Ballinger Publishing Company, 1989.

Bush, Vannevar. *Science the Endless Frontier.* New York: Arno Press, 1980.

Cabot, Thomas D. "Is American Education Competitive?" *Harvard Magazine.* Spring 1986: 42.

Calleo, David. *Beyond American Hegemony: The Future of the Western Alliance.* New York: Basic Books, 1987.

"Can the Electronics Industry Survive the Cold War?" *Electronic Business.* January 1993: 68.

Cantwell, Michael N. "Global Competition: U.S. Industry's Hidden Advantages." *Industry Week.* 7 October 1991: 53.

"The Case Against Free Trade." *Fortune.* 20 April 1992: 159.

Caves, Richard E., and Lawrence B. Krause. *Britain's Economic Performance.* Washington, D.C.: The Brookings Institution, 1980.

———. "International Trade and Industrial Organization: Problems Solved and Unsolved." *European Economic Review* 28 (August 1985): 377-395.

Celis, William. "Study Says Half of Adults in U.S. Can't Read or Handle Arithmetic." *New York Times.* 9 September 1993: A1.

Center for Strategic and International Studies. *Integrating Commercial and Military Technologies for National Strength.* Report of the CSIS Steering Committee on Security and Technology. Washington, D.C.: Government Printing Office, 1991.

Chandler, Alfred. *Scale and Scope: The Dynamics of Industrial Capitalism.* Cambridge: Harvard University Press, 1990.

"A Chip War With Japan?" *Business Week.* 22 June 1992: 34.

Choate, Patrick. *American Economic Power: Redefining National Security for the 1990's.* Report prepared for the Joint Economic Committee of the U.S. Congress, Washington, D. C.: Government Printing Office, November 1989.

Cohen, Linda R., and Roger G. Noll. *The Technology Pork Barrel.* Washington, D.C.: The Brookings Institution, 1991.

Cohen, Richard, and Peter Wilson. *Superpowers in Economic Decline: U.S. Strategy for the Transcentury Era.* New York: Crane Russak, 1990.

Cohen, Roger. "Europe State-Industry Ties: Success and Utter Failures." *New York Times.* 11 August 1992: A1.

Cohen, Stephen, and John Zysman. *Manufacturing Matters: The Myth of the Post-Industrial Economy.* New York: Basic Books, 1987.

Cole, Bernard. "Smart Memories are Eating into the Jelly Bean Market." *Electronics.*

5 February 1987: 27-29.

Committee on the Role of the Manufacturing Technology Program in the Defense Industrial Base, and the Commission on Engineering and Technical Systems. *Manufacturing Technology: Cornerstone of a Renewed Defense Industrial Base.* Washington, D.C.: Office of the Secretary of Defense, 1987.

Coover, H. W., "Programmed Innovation Strategy for Success." In *The Positive Sum Strategy: Harnessing Technology for Economic Growth*, Nathan Rosenberg and Ralph Landau, eds. Washington, D.C.: National Academy Press, 1986.

Crew, M., ed. *Competition and the Regulation of Industries.* Boston: Kluwer Academic, 1991.

Curien, Hubert. "The Revival of Europe." In *A High Technology Gap? Europe, America, and Japan*, Frank Press, Hubert Curien, and Keichi Oshima, eds. New York: Council on Foreign Relations, 1987.

"Current Conditions for Japan's Semiconductor Industry." *Japan 21st Century.* January 1993: 43.

Currie, Malcolm and others. *Defense Industrial Cooperation with Pacific Rim Nations.* Washington, D.C.: Office of Under Secretary of Defense for Acquisition, 1989.

Dataquest Incorporated and Quick, Finan, and Associates. *The Drive for Dominance: Strategic Options for Japan's Semiconductor Industry.* San Jose: Dataquest, 1988.

Deardorff, A.V. "Testing Trade Theories and Predicting Trade Flows." In *Handbook of International Economics*, R.W. Jones, ed. Amsterdam: Elsevier, 1985.

Degrasse, Robert. "The Military and Semiconductors." In *The Militarization of High Technology*, John Tirman, ed. Cambridge: Ballinger, 1984.

Denison, Edward. *The Sources of Economic Growth.* Washington, D.C.: Committee for Economic Development, 1962.

———. *Accounting for Slower Economic Growth: The United States in the 1970's.* Washington, D.C.: The Brookings Institution, 1979.

Derian, Jean-Claude. *America's Struggle for Leadership in Technology.* Severen Schaeffer, trans. Cambridge: MIT Press, 1990.

Dermer, Jerry, ed. *Competitiveness through Technology: What Business Needs From Government.* Toronto: Lexington Books, 1986.

Dertouzous, Michael L., Richard K. Lester, and Robert M. Solow. *Made in America: Regaining the Productivity Edge.* The MIT Commission on Industrial Productivity. Cambridge: MIT Press, 1989.

Deutsch, Edwin, and Wolfgang Schopp. *The Economics of Military Expenditures, Civil Versus Military R &D Expenditures and Industrial Productivity.* New York: St. Martins Press, 1987.

Developing World Industry and Technology, Inc. *Sources of Japan's International Competitiveness in the Consumer Electronics Industry: An Examination of Selected Issues.* Report submitted to U.S. Office of Technology Assessment. Washington, D.C.: Government Printing Office, June 1980.

Dinopoulos, Elias, and James Oehmke. "High Technology Industry Trade and Investment: The Role of Factor Endowments." *Journal of International Economics* 34 (February 1993): 471-513.

Diwan, Romesh, and Chandana Chakraborty, eds. *High Technology and International Competitiveness.* New York: Praeger, 1991.

Dixit, Avinash K. "The Use of Protection and Subsidies for Entry Promotion and

Deterrence." *American Economic Review* 75 (March 1985): 129-156.

———. "Trade Policy: An Agenda for Research." In *Strategic Trade Policy and the New International Economics*, Paul Krugman, ed. Cambridge: MIT Press, 1986.

Dixit, Avinash K., and Gene M. Grossman. "Targeted Export Promotion with Several Oligopolistic Industries." *Journal of International Economics* 21 (November 1986): 233-249.

Dollar, David, and Edward Wolff. "Convergence of Industry Labor Productivity Among Advanced Economies,1936-1982." *Review of Economic Statistics* 70 (November 1988): 549-558.

Dosi, Giovanni. "Semiconductors: Europe's Precarious Survival in High Technology." In *Europe's Industries: Public and Private Strategies for Change*, Geoffrey Shepard, ed. Ithaca: Cornell University Press, 1983.

———. "Finance, Innovation and Industrial Change." *Journal of Economic Behavior & Organization* 13 (June 1990): 10-16.

Duke, Simon. *The Burdensharing Debate: A Reassessment*. London: Macmillan, 1993.

Eaton, Johnathon, and Gene M. Grossman. "Optimal Trade and Industrial Policy Under Oligopoly." *Quarterly Journal of Economics* 101 (May 1986): 383-406.

"Economic and Financial Indicators." *The Economist*. 26 August 1993.

"Economics Focus: America's Twin Deficits." *The Economist*. 12 October 1988: 69.

"Economics Focus: National Subsidies and the EEC's Single Market." *The Economist*. 30 January 1988: 56.

"Economics of Managed Trade." *The Economist*. 22 September 1990: 17-22.

Eisenger, Peter. *Rise of the Entrepreneurial State: State and Local Economic Development Policy in the United States*. Madison: University of Wisconsin Press, 1986.

Eismeier, Theodore. "The Case Against Industrial Policy." *Journal of Contemporary Studies* 6 (Spring 1983): 145-156.

Ellison, John, Jeffrey Frumkin, and Timothy Stanley. *Mobilizing U.S. Industry: A Vanishing Option for National Security?* Boulder: Westview Press, 1988.

Erickson, Keith, and Ashok Kanagal. "Partnering for Total Quality." *Quality*. September 1992: 16-20.

ESPRIT Review Board. *The Review of ESPRIT 1984-1988*. Luxembourg: Commission of the European Community, May 1989.

Ettlie, John. "Manufacturing Productivity by the Numbers." *Production* 104 (December 1992): 12-13.

European Communities Commission (DG IV). Official Journal of the European Communities. no. L33 Luxembourg: EC Commission, February 4, 1989.

———. *Industrial Consequences of Targeting*. Luxembourg: Commission of the European Community, January 1990.

———. *The European Electronics and Information Technology Industry: State of Play, Issues at Stake, and Proposals for Action*: DG XIII Telecommunications, Information Industries, and Innovation. Luxembourg: Commission of the European Communities, March 1991.

Fagerberg, Jan. "Why Growth Rates Differ." In *Technical Change and Economic Theory*, Giovanni Dosi, Christopher Freeman, Luc Soete, and Richard Nelson, eds. New York: Francis Pinter, 1988.

Finan, William F., and Chris Amundsen. *Analysis of the Relative Economic Benefits of Tax Depreciation Policies for Semiconductor Equipment and Facilities in the*

United States and Japan. San Francisco: Technecon Analytic Research Inc., February 1992.

Flamm, Kenneth. "Technology Policy in International Perspective." In *Policies for Industrial Growth in a Competitive World.* Report prepared for the Joint Economic Committee of the 99th U.S. Congress., 2d Session, 29-39. Washington, D.C.: Government Printing Office, 1984.

———. *Targeting the Computer: Government Support and International Competition.* Washington, D.C.: The Brookings Institution, 1987.

———. "Semiconductors." In *Europe 1992: An American Perspective*, Gary Clyde, Hufbauer, ed. Washington, D.C.: The Brookings Institution, 1990.

———. "Making New Rules: High Tech Trade Friction and the Semiconductor Industry." *The Brookings Review* 9 (Spring 1991): 22-29.

Forester, Thomas. *Silicon Samurai: How Japan Conquered the World's Information Technology Industry.* Cambridge: Blackwell, 1993.

Fransman, Martin. *Technology and Economic Development.* Boulder: Westview Press, 1986.

———. *The Market and Beyond: Cooperation and Competition in Information Technology Development in the Japanese System.* Cambridge: Cambridge University Press, 1990.

Freeman, Christopher. *Technology Policy and Economic Performance: Lessons From Japan.* London: Francis Pinter, 1987.

Freeman, Christopher, and Margaret Sharp, eds. *Technology and the Future of Europe: Global Competition and the Environment in the 1990's.* London: Francis Pinter, 1991.

Fujiwara, Okuno. "Interdependence of Industries: Coordination Failure and Strategic Promotion of an Industry." *Journal of International Economics* 25 (1988): 87-102.

Gee, Sherman. *Technology Transfer, Innovation, and International Competitiveness.* New York: John Wiley, 1981.

Georghio, L. G., and J. S. Metcalfe. "Evaluation of the Impact of European Community Research Programmes Upon Industrial Competitiveness." *R&D Management* 23 (1993): 161-169.

Ghemawat, Pankaj. "Sustainable Advantage." *Harvard Business Review* 64 (September/October 1986): 179-193.

Gilpin, Robert. *War and Change in World Politics.* New York: Cambridge University Press, 1981.

Giovanni, Dosi, Christopher Freeman, Richard Nelson, and Luc Soete, eds. *Technical Change and Economic Theory.* New York: Pinter Publishers, 1988.

Glimm, James. *Mathematical Sciences, Technology, and Economic Competitiveness.* Washington D.C.: National Academy Press, 1991.

Goodman, Richard, and Julian Pavon, eds. *Planning for National Technology Policy.* New York: Praeger Special Studies, 1984.

Graham, Otis L. Jr. *Losing Time: The Industrial Policy Debate.* Cambridge: Harvard University Press, 1992.

Gresser, Julian. *High Technology and Japanese Industrial Policy: A Strategy for U.S. Policymakers.* Report prepared for the subcommittee on Trade and Committee on Ways and Means, U.S. House of Representatives. Washington, D.C.: Government Printing Office, 1980.

Griliches, Zvi. "Research Expenditures and Growth Accounting." In *Science and Technology in Economic Growth*, B.R. Williams, ed. New York: John Wiley, 1973.

———. "Issues in Assessing the Contribution of R&D to Productivity Growth." *Bell Journal of Economics* 10 (April 1979): 92-116.

Gross, Neil. "Inside Hitachi." *Business Week.* 28 September 1992: 92-100.

Grossman, Gene M. "Strategic Export Promotion: A Critique." In *Strategic Trade Policy and the New International Economics*, Paul Krugman, ed. Cambridge: MIT Press, 1986.

Grossman, Gene M., and J. David Richardson. *Strategic Trade Policy: A Survey of Issues and Early Analysis.* Special Papers in International Economics no. 15. Princeton: International Finance Section, 1985.

Grossman, Gene M., and Elhanan Helpman. *Comparative Advantage and Long-Run Growth.* Cambridge: National Bureau of Economic Research, 1989.

———. *Innovation and Growth in the Global Economy.* Cambridge: MIT Press, 1991.

Gsand, William. Personal interview at Hitachi America LTD. 12 August 1992.

Guerrieri, Paolo, and Carlo Milana. "Technological and Trade Competition in High Tech Products." *BRIE Working Papers* 54 Berkeley: University of California Berkeley, 1991.

Hafner, Katie. "Does Industrial Policy Work? Lessons From Sematech." *New York Times.* 7 November 1993: F3.

Hall, Peter, ed. *Technology, Innovation, and Economic Policy.* New York: St. Martin's Press, 1986.

Harris, Richard. *Trade, Industrial Policy, and International Competition.* Toronto: University of Toronto Press, 1985.

Hatsopoulos, George N. "High Cost of Capital: Handicap of American Business." In *U.S Department of Commerce Calculations*, U.S. Department of Commerce. Washington, D.C.: Government Printing Office, April 1983.

Hatsopoulos, George N., Paul R. Krugman, and Lawrence H. Summers. "U.S. Competitiveness: Beyond the Trade Deficit." *Science* 241 (1988): 299-307.

Haw, Kenneth, and Adam Jaffe. "Effects of Liquidity on Firms' R&D Spending." *Economics of Innovation and New Technology* 2 (1993): 275-282.

Hayes, Robert, and Jaikumar Rachandran. "Manufacturing's Crisis: New Technologies, Obsolete Organizations." *Harvard Business Review* 66 (September/October 1988): 77-85.

Heiduk, Gunter, and Kozo Yamamura, eds. *Technological Competition and Interdependence: The Search for Policy in the United States, West Germany, and Japan.* Seattle: University of Washington Press, 1990.

Helphman, Elhanan, and Paul Krugman. *Market Structure and Foreign Trade: Increasing Returns, Imperfect Competition, and the International Economy.* Cambridge: MIT Press, 1985.

Henderson, Jeffrey. *The Globalization of High Technology Production: Society, Space, and Semiconductors in the Restructuring of the Modern World.* New York: Routledge, 1989.

"Here Comes Erasable Optical Discs." *The Economist.* 5 December 1990: 88.

Hilke, John C. *International Competitiveness and the Trade Deficit.* Washington, D.C.: Federal Trade Commission, 1987.

Hill, Graham. *European Industrial Policy.* New York: St. Martin's Press, 1986.

Hilts, Philip. "Research Pact Under Fire." *New York Times*. 3 December 1993: A16.

Hindley, Brian. *State Investment Companies in Western Europe: Picking Winners or Backing Losers?* New York: St. Martin's Press, 1983.

Hippell, Eric Von. *Sources of Innovation.* Oxford: Oxford University Press, 1988.

Hobday, Mike. "The European Semiconductor Industry: Resurgence and Rationalization." In *Technology and the Future of Europe: Global Competition and the Environment in the 1990's*, Christopher Freeman and Margaret Sharp, eds. London: Francis Pinter, 1991.

Hof, Robert D. "Inside Intel." *Business Week*. 1 June 1992: 86-95.

Holberton, Simon. "The Differing Costs of Capital." *The Financial Times*. 1 June 1990: 16.

Holland, Max. *When the Machine Stopped.* Cambridge: Harvard Business School Press, 1989.

Horiuchi, Akiyoshi. "Influence of the Japan Development Bank Loans on Corporate Investment Behavior." *Journal of the Japanese and International Economy* 7 (4) (December 1993): 441-465.

"How Boeing Fights Airbus." *The Economist*. 30 January 1988: 51.

Howell, Thomas R., Brent L. Bartlett, and Warren Davis. *Creating Advantage: Semiconductors and Government Industrial Policy in the 1990's*. San Francisco: Semiconductor Industry Association, 1992.

Hu, Yao-su. "Global or Stateless Corporations are National Firms with International Operations." *California Management Review* 2 (Winter 1992): 107-123.

Imai, Ken-ichi. "Japan's Industrial Policy for High Technology Industry." In *Japan's High Technology Industries: Lessons and Limitations of Industrial Policy*, Hugh Patrick, ed. Seattle: University of Washington Press, 1986.

"Inside Intel." *Business Week*. 1 June 1992: 86.

International Monetary Fund. *International Financial Statistics Yearbook.* Washington, D.C.: International Monetary Fund, 1988.

———. *International Financial Statistics Yearbook.* Washington, D.C.: International Monetary Fund, 1991.

Ishihara, Shintaro. *The Japan That Can Say No*. Frank Baldwin, trans. New York: Simon & Schuster, 1991.

Itoh, Motoshige, and Kazuharu Kiyono, eds. *Economic Analysis of Industrial Policy*. New York: Academic Press, 1991.

Itoh, Takatoshi. *The Japanese Economy*. Cambridge: MIT Press, 1992.

Jacquemin, Alex. *The New Industrial Organization: Market Forces and Strategic Behavior*. Cambridge: MIT Press, 1987.

Japan Ministry of International Trade and Industry. *Vision of MITI Policies in the 1980*'s. Tokyo: MITI, 1980.

———. *Vision of MITI Policies in the 1990's*. Tokyo: MITI, 1990.

"Japan Trade Barriers Lift Costs Sharply." *Wall Street Journal*. 15 December 1994.

"Japanese Finance: End of an Era." *The Economist*. 12 October 1988: 58.

JESSI Planning Committee. *Jessi Program—Results of the Planning Phase*. Luxembourg: Commission of the European Communities, February 1990.

Johansen, Robert C. *Toward an Alternative Security System*. New York: World Policy Institute, 1983.

Johnson, Chalmers. *MITI and the Japanese Miracle: The Growth of Industrial Policy, 1925-1975*. Stanford: Stanford University Press, 1982.

———. "Keiretsu: An Outsiders View." *International Economic Insights* (September-October 1990): 16-20.

———. ed. *The Industrial Policy Debate.* San Francisco: ICS Press, 1984.

Johnson, Harry G. "Optimum Tariffs and Retaliation." *Review of Economic Studies* 21 (1953): 142-153.

Kanz, John W. "Technology, Globalization, and Defense: Military Electronic Strategies in a Changing World." *International Journal of Technology Management* 8 (1993): 59-76.

Karmin, Monroe W. "Industrial Policy: What is it? Do We Need One?" *U.S. News and World Report.* 3 October 1983: 23.

Kennedy, Paul. *The Fall and Rise of Nations: The Future of America.* Washington D.C.: Woodrow Wilson International Center for Scholars, 1987.

———. "The (Relative) Decline of America." *The Atlantic Monthly.* August 1987: 29-41.

———. *The Rise and Fall of the Great Powers.* New York: Random House, 1987.

Kierzowski, Henry, ed. *Monopolistic Competition and International Trade.* Oxford: Clarendon Press, 1984.

Kindleberger, Charles. *American Business Abroad.* New Haven: Yale University Press, 1969.

———. "Dominance and Leadership in the International Economy." *International Studies Quarterly* 25, no. 2 (June 1981): 242-254.

Knorr, Klaus, "Economic Interdependence and National Security." In *Economic Issues and National Security*, Klaus Knorr, ed. Lawrence: Allen Press, 1977.

———. ed. *Economic Issues and National Security.* Lawrence: Allen Press, 1977.

Komiya, Ryutaro, and Masahiro Okuno, eds. *Industrial Policy of Japan.* New York: Academic Press, 1988.

Kopcke, Richard, and Mark Howrey. "A Panel Study of Investment: Sales, Cash Flow, the Cost of Capital, and Leverage. *New England Economic Review* (January/February 1994): 9-30.

Krauss, Melvyn. "Europeanizing the U.S. Economy: The Enduring Appeal of the Corporatist State." In *The Industrial Policy Debate*, Chalmers Johnson, ed. San Francisco: Institute for Contemporary Studies, 1984.

Krouse, Clement. "Competitive Advantage, Cost of Capital and the Financial Relationship System in Japan." *Journal of Institutional and Theoretical Economics* 149 (4) (December 1993): 634-655.

Krueger, Anne O. "Theory and Practice of Commercial Policy: 1945-1990." *NBER Working Papers* 3569. Cambridge: National Bureau of Economic Research, December 1990.

Krugman, Paul. "Import Protection as Export Promotion: International Competition in the Presence of Oligopoly and Economies of Scale." In *Monopolistic Competition and International Trade*, Henry Kierzowski, ed. Oxford: Clarendon Press, 1984.

———. "Market Access and International Competition: A Simulation Study of 16K Random Access Memory." In *Empirical Methods for International Trade*, Robert Feenstra, ed. Cambridge: MIT Press, 1985.

———. "New Thinking About Trade Policy." In *Strategic Trade Policy and the New International Economics*, Paul Krugman, ed. Cambridge: MIT Press, 1986.

———. "Is Free Trade Passé?" *Economic Perspectives* 1 (Fall 1987): 36-44.

———. "Strategic Sectors and International Competition." In *U.S. Trade Policies ina Changing World Economy*, Robert Stern, ed. Cambridge: MIT Press, 1987.

———. "Has the Adjustment Process Worked?" *Policy Analyses in International Economics* 34. Washington, D.C.: Institute for International Economics, 1990.

———. "Technology and International Competition: Overview." Paper prepared for a National Academy of Engineering Symposium on *Linking Trade and Technology Policies: An International Comparison.* Washington, D. C.: National Academy of Sciences, 1991.

———. *The Age of Diminished Expectations: U.S. Economic Policy in the 1990's.* Cambridge: MIT Press, 1992.

———. "Competitiveness: A Dangerous Obsession." *Foreign Affairs.* March 1994: 27-39.

———. ed. *Strategic Trade Policy and the New International Economics.* Cambridge: MIT Press, 1986.

Kuttner, Robert. *The End of Laissez-Faire: National Purpose and the Global Economy After the Cold War.* New York: Alfred A. Knopf, 1991.

Lamm, Richard. *Megatraumas: America at the Year 2000.* Boston: Houghton Mifflin, 1985.

Lawrence, Robert Z. "Before Industrial Policy." *New York Times.* 30 November 1983: A31.

———. *Can America Compete?* Washington, D.C.: The Brookings Institution, 1984.

Lazonick, William. *Competitive Advantage on the Shop Floor.* Cambridge: Harvard University Press, 1990.

Le, Can. "The Role of R&D in High Technology Trade: An Empirical Analysis." *The Journal of Economics* 14 (1988): 97-109.

Lee, Michelle K. "High Technology Consortia: A Panacea for America's Technological Competitiveness Problems?" *High Technology Law Journal* 6:2 (1992): 335-362.

Levine, Johnathon. "European High Technology Tries to do Something Drastic: Grow Up." *Business Week.* 25 March 1991: 48.

Levinson, Phyllis. *The Federal Entrepreneur: The Nation's Implicit Industrial Policy.* Washington, D.C.: Urban Institute, 1982.

Libicki, Martin C. *What Makes Industries Stategic.* Washington, D.C.: National Defense University, November 1989.

Lodge, George. *Perestroika for America: Restructuring U.S. Business-Government Relations for Competitiveness in the World Economy.* Boston: Harvard Business School Press, 1990.

Lohr, Steve. "Shrinking Payrolls." *New York Times.* 17 December 1992: A1.

Magaziner, Ira. "Growing Our Economy." *Business & Economics Review,* Vol. 39 (April/June 1993): 8-12.

Magaziner, Ira, and Thomas Trout. *Japanese Industrial Policy.* Berkeley: Institute of International Studies, 1981.

Malerba, Franco. *The Semiconductor Business: The Economics of Rapid Growth and Decline.* London: Frances Pinter, 1985.

Mansfield, Edwin. *The Speed and Cost of Industrial Innovation in Japan and the United States.* University of Pennsylvania Working Paper. Philadelphia: University of Pennsylvania, 1985.

Markoff, John. "Unable to Beat Them, I.B.M. Now Joins Them." *New York Times.* 6 July 1992: D1.

Marston, Richard C. "Price Behavior in Japanese and U.S. Manufacturing." *National Bureau of Economic Research* Working Paper no. 3364. Cambridge: National Bureau of Economic Research, 1990.

May, Ernest R. "The U.S. Government: a Legacy of the Cold War." *Journal of Diplomatic History* 16 (Spring 1992): 260-277.

McCauland, Richard. "Matushita Shows 16M DRAM Prototype." *Electronic News*. 21 May 1990: 23.

McCauley, Robert, and Steven Zimmer. *Explanations for International Differences in the Cost of Capital*. New York: Federal Reserve Bank of New York, 1989.

Methe, David T. *Technological Competition in Global Industries: Marketing and Planning Strategies for American Industry*. New York: Quorom Books, 1991.

Mishel, Lawrence, and David Frankel, *The State of Working America*. Washington, D.C.: Economic Policy Institute, 1990.

Moran, Theodore H. "The Globalization of America's Defense Industries: Managing the Threat of Foreign Dependence." *International Security* 15 (Summer 1990): 57-100.

Morris, Peter. *A History of the World Semiconductor Industry*. London: Frances Pereginus on behalf of the Institution of Electrical Engineers, 1990.

Mowery, David. *Science and Technology Policy in Interdependent Economies*. Cambridge: Cambridge University Press, 1994.

Mowery, David, and Nathan Rosenberg. *Technology and the Pursuit of Economic Growth*. Cambridge: Cambridge University Press, 1989.

Nelson, Richard. *High Technology Policies: A Five Nation Comparison*. Washington, D.C.: American Enterprise Institute, 1985.

Nelson, Richard, and George Eads. "Japanese High Techology Policy: What Lessons for the U.S.? In *Japan's High Technology Industries: Lessons and Limitations of Industrial Policy*, Hugh Patrick, ed. Seattle: University of Washington Press, 1986.

Nelson, Richard, and Gavin Wright. "The Rise and Fall of American Technological Leadership: The Postwar Era in Historical Perspective." *Journal of Economic Literature* 30 (December 1992): 1931-1964.

Niosi, Jorge, ed. *Technology and National Competitiveness*. Montreal: McGill-Queen's University Press, 1991.

Noguchi, Yukio. "The Government Business Relationship in Japan." In *Policy and Trade Issues of the Japanese Economy*, Kozo Yamamura, ed. Seattle: University of Washington Press, 1982.

Ohmae, Kenichi. *The Borderless World*. London: Collins, 1990.

Okimoto, Daniel. *Between MITI and the Market: Japanese Industrial Policy for High Technology*. Stanford: Stanford University Press, 1989.

Okimoto, Daniel, Takuo Sugano, and Franklin Weinstein. *Competitive Edge: The Semiconductor Industry in the U.S. and Japan*. Stanford: Stanford University Press, 1984.

Organization for Economic Cooperation and Development. *Economic Outlook*. Paris: OECD, 1975.

———. *National Accounts: Main Aggregates* Vol. 1. Brussels: OECD, 1984.

———. *Trade in High-Technology Products: the Semiconductor Industry*. Paris: OECD, 1984.

———. *Handbook of Economic Statistics* Paris: OECD, 1986.

———. *Economic Outlook*. Vol. 47. Paris: OECD, 1990.

———. *Industrial Policy in OECD Countries*. Paris: OECD, 1990.

———. *Handbook of Economic Statistics.* Paris: OECD, 1991.

———. *Strategic Industries ina Global Economy: Policy Issues for the 1990's.* Paris: OECD, 1991.

———. *Structural Competitiveness and National Systems of Production.* Paris: OECD, 1991.

———. *Main Economic Indicators.* Paris: OECD, April 1994.

Ostry, Sylvia. "Governments and Corporations in a Shrinking World: Trade & Innovation Policies in the United States, Europe, and Japan." *Columbia Journal of World Business* 25 (Spring/Summer 1990): 10-16.

"Overwhelmed by Capital Investment." *The Economist.* 9 September 1990: 71.

Ozaki, Robert. "How the Japanese Industrial Policy Works." In *The Industrial Policy Debate*, Chalmers Johnson, ed. San Francisco: Institute of Contemporary Studies, 1984.

Parsons, Carol A. "The Changing Shape of Domestic Employment in High-Tech Industry: The Case of International Trade in Semiconductors." In *The Dynamics of Trade and Employment*, Laura D'Andrea Tyson, ed. Cambridge: Ballinger Publishing Company, 1988.

Pascall, Glenn R., and Robert D. Lamson. *Beyond Guns & Butter: Recapturing America's Economic Momentum After a Military Decade.* New York: Brassey's Inc., 1991.

Pascall, Richard T., and Anthony G. Athos. *The Art of Japanese Management: Applications for American Executives.* New York: Warner Books, 1982.

Patrick, Hugh, ed. *Japan's High Technology Industries: Lessons and Limitations of Industrial Policy.* Seattle: University of Washington Press, 1986.

———. "Japanese High Technology Industrial Policy in Comparative Context." In *Japan's High Technology Industries: Lessons and Limitations of Industrial Policy,* Hugh Patrick, ed. Seattle: University of Washington Press, 1986.

"Pentagon's Push to Bolster Competitiveness." *Challenges.* September 1988.

"Personal Savings Rise." *The Japan Economic Journal.* 15 September 1990: 89.

Piore, Michael, and Charles Sabel. *The Second Industrial Divide.* New York: Basic Books, 1984.

Pollack, Andrew. "Intel Adds Cheaper Chip at Top of the Line." *New York Times.* 22 April 1991: D4.

———."I.B.M.-Toshiba Joint Chip Venture" *New York Times.* 22 June 1992: D3.

———."A Chip Powerhouse is Challenged." *New York Times.* 17 December 1992: D1.

Porter, Michael. T*he Competitive Advantage of Nations.* New York: The Free Press, 1990.

Prestowitz, Clyde V. *Trading Places: How We Allowed Japan to Take the Lead.* New York: Basic Books, 1988.

———. "In a Single Decade." In *Powernomics: Economics and Strategy After the Cold War,* Clyde V. Prestowitz, ed. Lanham: Madison Books, 1991.

———. *Powernomics: Economics and Strategy After the Cold War.* Lanham: Madison Books, 1991.

Protzman, Ferdinand. "Why a Lower Dollar Didn't Work." *New York Times.* 1 December 1992: D1.

Prowse, Michael. "Is America in Decline." *Harvard Business Review* 70 (July/August 1992): 34-45.

Quick, Finan, and Associates. *International Transfer of Semiconductor Technology through U.S. Based Firms.* NBER Working Paper 118. New York: National Bureau of Economic Research, 1975.

———. *The Drive for Dominance: Strategic Options for Japan's Semiconductor Industry.* San Jose: Dataquest, 1988.

Raleigh News and Observer. 17 August 1983.

Rasmussen, David, and Larry Ledebur. "The Role of State Economic Development Programs in National Industrial Policy." In *Revitalizing the U.S. Economy,* Stevens Redburns, Terry Buss, and Larry Ledebur, eds. New York: Praeger, 1986.

"R&D Scoreboard." *Business Week.* 15 June 1990: 86.

Reich, Robert. "Why the U.S. Needs an Industrial Policy." *Harvard Business Review* (January/February 1982): 107-119.

———. "The Real Economy." *The Atlantic Monthly.* February 1991: 35-52.

———. *The Work of Nations: Preparing Ourselves for 21st Century Capitalism.* New York: A. A. Knopf, 1991.

Reich, Robert, and Ira Magaziner. *Minding America's Business.* New York: Harcourt Brace Jovanovich, 1983.

Ricardo, David. *On the Principles of Political Economy and Taxation.* London: John Murray, 1817.

Richards, Evelyn. "Chief Sees New Aim for MCC: Consortium Pushed to Businesslike." *San Jose Mercury News.* 5 May 1991: E1.

Richardson, David. "The Political Economy of Strategic Trade Policy." *International Organization* 44 (Winter 1990): 106-132.

———. "New Trade Theory and Policy a Decade Old: Assessment in a Pacific Context." Paper Presented at the First Australian Fulbright Symposium on *Managing International Relations in the 1980's.* Canberra, March 1992.

Ristelhueber, Robert. "Another Pleasant Surprise This Year for Chip Makers." *Electronic Business.* January 1993: 95-96.

———. "U.S. Makers Climb to World's Top 15." *Electronic Business.* January 1993: 17.

———. "Setting Sun?" *Electronic Business.* April 1994: 52.

Rosenbaum, David E. "Promises Before Policies." *New York Times.* 12 November 1992, A1.

Rosenberg, Nathan. *Inside the Black Box: Technology and Economics.* Cambridge: Cambridge University Press, 1982.

Rosenberg, Nathan, and Ralph Landau. *The Positive Sum Strategy: Harnessing Technology for Economic Growth.* Washington, D.C.: National Academy Press, 1986.

Rosenberg, Nathan, and David Mowery. *Technology and the Pursuit of Economic Growth.* Cambridge: Cambridge University Press, 1990.

Rosencrance, Richard. *America's Economic Resurgence.* New York: Harper & Row Publishers, 1990.

Ross, Andrew L., ed. *The Political Economy of Defense: Issues and Perspectives.* New York: Greenwood Press, 1991.

Salvatore, Dominic. *The Japanese Trade Challenge and the U.S. Response.* Washington, D.C.: Economic Policy Institute, 1990.

Samuels, Richard J., and Benjamin C. Whipple. "The FSX and Japan's Strategy in Aerospace." *Technology Review* 92 (October 1989): 42-51.

Sanders, Ralph. *International Dynamics of Technology.* London: Greenwood Press, 1983.

Sanquini, Richard. Personal interview at National Semiconductor. 11 August 1992.

Savitz, Eric. "Time for a Change: Semiconductor Equipment Makers Poised for Turnaround." *Barrons.* 16 November 1992: 35.

Scherer, Frederic M. "Inter-Industry Technology Flows and Productivity Growth." *Review of Economics and Statistics* 64 (November 1982): 627-634.

Schmidt, Christian, ed. *The Economics of Military Expenditures: Military Expenditures, Economic Growth and Fluctuations.* Proceedings of an International Economic Association Conference (Paris France). New York: St. Martin's Press, 1987.

Schmidt, Roland W. "Successful Corporate R&D." *Harvard Business Review* 63 May/June 1985): 124-128.

Schultze, Charles L. "Industrial Policy: A Solution in Search of a Problem." *California Management Review* 24 (Summer 1983): 84-102.

Schumpeter, Joseph. *Business Cycles: A Theoretical, Historical, and Statistical Analysis of the Capitalist Process.* New York: McGraw Hill, 1939.

———. *Capitalism, Socialism, and Democracy.* New York: Harper Brothers, 1942.

Scott, Bruce."How Practical is National Economic Planning." *Harvard Business Review* (March/April 1978): 156-164.

———. "National Strategies: Key to International Competition." In *U.S. Competitiveness in the World Economy*, Bruce Scott and George Lodge, eds. New York: Norton, 1987.

SEMATECH. Annual Report. 1991.

Semiconductor Industry Association. *The Effect of Government Targeting on World Semiconductor Competition.* Cupertino: Semiconductor Industry Association, 1983.

———. *Four Years of Experience Under the U.S.-Japan Semiconductor Agreement: Fourth Annual Report to the President.* Cupertino: Semiconductor Industry Association, 1990.

Sen, Gautam. *The Military Origins of Industrialization and International Trade Rivalry.* New York: St. Martin's Press, 1984.

Sharp, Margaret. *European Technological Collaboration.* Chatham House Papers no. 36. London: Royal Institute of International Affairs, 1987.

———. "The Single European Market and European Technology Policies." In *Technology and the Future of Europe: Global Competition and the Environment in the 1990's*, Christopher Freeman and Margaret Sharp, eds. London: Francis Pinter, 1991.

Shepard, Geoffrey. *Europe's Industries: Public and Private Strategies for Change.* Ithaca: Cornell University Press, 1983.

Shultz, Richard, and Robert Pfaltzgraff, eds. *U.S. Defense Policy in an Era of Constrained Resources.* Toronto: Lexington Books, 1990.

Sigurdson, Jon. *Industry and State Partnership in Japan: The Very Large Scale Integration Project.* Lund: Research Policy Institute, 1986.

Soete, Luc. "Technological Change and International Trade." In *Technical Change and Economic Theory*, Giovanni Dosi, Christopher Freeman, Luc Soete, and Richard Nelson, eds. New York: Francis Pinter, 1988.

"Software Copyright." *The Economist.* 9 September 1990: 74.

Spar, Deborah, and Raymond Vernon. *Beyond Globalism: Remaking American Foreign Economic Policy.* New York: The Free Press, 1989.

Spence, Michael A. "Cost Reduction, Competition, and Industry Performance." *Econometrica* 52 (1984): 101-121.

Spencer, Barbara. "What Should Strategic Trade Policy Target?" In *Strategic Trade Policy and the New International Economics*, Paul Krugman, ed. Cambridge: MIT

Press, 1986.

Stackleberg, Heinrich Von. *The Theory of the Market Economy.* Alan T. Peacock, trans. London: W. Hodge, 1952.

Standard & Poor's. "Electronics." *Standard & Poor's Industry Survey.* June 1994.

Stein, Herbert. "Don't Fall for Industrial Policy." *Fortune.* 14 November 1983: 78.

———. *Presidential Economics.* New York: Simon & Schuster, 1984.

Stern, Robert, ed. *U.S. Trade Policies in a Changing World Economy.* Cambridge: MIT Press, 1987.

Sterngold, James. "A Frenzy Over Spoils." *New York Times.* 25 October 1992: 10.

Stewart, William L. *Problems of Small High Technology Firms.* Special Report of the National Science Foundation 81-305. Washington, D.C.: National Science Foundation, 1984.

Tassey, Gregory. *Technology Infrastructure and Competitive Position.* Norwell: Kluwer Academic Publishers, 1992.

Taylor, Jared. *Shadows of the Rising Sun: A Critical View of the Japanese Miracle.* New York: Morrow, 1983.

Teece, David J. "Foreign Investment and Technological Development in Silicon Valley." *California Management Review* (Winter 1992): 88-105.

Thurow, Lester C. *The Zero Sum Society: Distribution and the Possibilities of Economic Change.* New York: Basic Books, 1980.

———. *The Case for Industrial Policies.* Washington, D.C.: Center for National Policy, 1984.

———. *Toward a High-Wage, High Productivity Service Sector.* Washington, D.C.: Economic Policy Institute, 1989.

———. *Head to Head: The Coming Economic Battle Among Japan, Europe, and America.* New York: William Morrow, 1992.

Tirman, John, ed. *The Militarization of High Technology.* Cambridge: Ballinger Publishing Company, 1984.

Tolley, George, James Hodge, and James Oehmke, eds. *The Economics of R&D Policy.* New York: Praeger Special Studies, 1985.

Trezize, Phillip H. "Industrial Policy is not the Major Reason for Japan's Success." *Brookings Review* 2 (Spring 1983): 13-21.

———. "Japan, the Enemy?" *Brookings Review* 8 (Winter 1989/1990): 3-13.

Trout, Thomas, and James Harf, eds. *National Security Affairs: Theoretical Perspectives and Contemporary Issues.* London: Transaction Books, 1982.

Tsuruta, Toshimasa. "The Myth of Japan, Inc." *Technology Review* 86 (July 1983): 22-25.

Tyson, Laura D'Andrea. *Who's Bashing Whom: Trade Conflict in High Technology Industries.* Washington, D.C.: Institute for International Economics, 1992.

Tyson, Laura D'Andrea, and William T. Dickens, eds. *The Dynamics of Trade and Employment.* Cambridge: Ballinger Publishing Company, 1988.

Uddis, Bernard. *The Challenge to European Industrial Policy: Impacts of Redirected Military Spending.* London: Westview Press, 1987.

U.S. Bureau of the Census. *Historical Statistics of the United States: Colonial Times to 1957.* Washington, D.C.: Government Printing Office, 1960.

U.S. Bureau of Labor Statistics, and U.S. Department of Labor. *Employment and Earnings.* Washington, D.C.: Bureau of Labor Statistics, March 1992.

U.S. Congressional Budget Office. *Federal Support for R&D and Innovation.* Washington, D.C.: Government Printing Office, 1984.

———. *Federal Financial Support for High Technology Industries*. Washington, D.C.: Government Printing Office, June 1985.

———. *Using R&D Consortia for Commercial Innovation: SEMATECH, X-Ray Lithography, and High-Resolution Systems*. Washington, D.C.: Government Printing Office, 1990.

U.S. Council on Competitiveness. *Gaining New Ground: Technology Priorities for America*. Washington, D.C.: Government Printing Office, 1991.

U.S. Congress, House of Representatives Committee on Science and Technology. *Japanese Technological Advances and Possible United States Responses Using Research Joint Ventures*. Hearings Before the Subcommittee on Investigations and Oversight and the Committee on Science and Technology. Washington, D.C.: Government Printing Office, June 1983.

U.S. Congress, Joint Economic Committee. *The 1981 Economic Report to the President*. Washington, D.C.: Government Printing Office, 1981.

———. *The 1991 Economic Report to the President*. Washington, D.C.:, Government Printing Office, 1991.

U.S. Defense Science Board to the Department of Defense. *Report of the Defense Science Board Task Force on Defense Semiconductor Dependency*. Washington, D.C.: Office of the Under Secretary of Defense for Acquisition, February 1987.

———. *Foreign Ownership and Control of US Industry*. Washington, D.C.: Government Printing Office, 1991.

U.S. Department of Commerce. *Report on the US Semiconductor Industry*. Washington, D.C.: Government Printing Office, 1979.

———. "Gross Product by Industry: Comments on Recent Criticisms." *Survey of Current Business* (July 1988): 132-38.

———. *Annual Survey of Manufacturers*. Washington, D.C.: Government Printing Office, 1989.

———. *Emerging Technologies*. Washington, D.C.: Government Printing Office, 1990.

U.S. Department of Defense. *Critical Technologies Plan for the Committees on Armed Services*. Washington, D.C.: Government Printing Office, 1989.

U.S. Federal Trade Commission. *64K DRAM Components from Japan, No. 731-TA-270*. Washington, D.C.: International Trade Commission, 1985.

———. *Dynamic Random Access Memory Semiconductors from Japan, No. 731-TA-300*. Washington, D.C.: International Trade Commission, 1986.

———. *Staff Report on the Semiconductor Industry*. Washington, D.C.: Government Printing Office, 1989.

U.S. General Accounting Office. *Industrial Policy: Case Studies in the Japanese Experience*. Washington, D.C.: Government Printing Office, 1982.

———. Foreign Industrial Targeting: U.S. Trade Law Remedies. Report to the Congress of the United States by the Comptroller General of the United States. Washington D.C.: General Accounting Office, May 1985.

———. *Federal Research: SEMATECH's Efforts to Strengthen the U.S. Semiconductor Industry*. Washington, D.C.: Government Printing Office, 1991.

———. *U.S. Business Access to Certain Foreign State-of-the-Art Technology*. Washington, D.C.: Government Printing Office, 1991.

U.S. International Trade Administration, U.S. Department of Commerce. *An Assessment of U.S. Competitiveness in High Technology Industries*. Washington, D.C.: Department of Commerce, February 1983.

U.S. International Trade Commission. *Foreign Industrial Targeting and Its Effects on U.S. Industries, Phase I: Japan.* USITC Publication no. 1437. Washington, D.C.: International Trade Commission, 1983.

———. *Foreign Industrial Targeting and Its Effects on U.S. Industries, Phase II: Europe.* USITC Publication 1437. Washington, D.C.: International Trade Commission, 1983.

———. *U.S. Competitiveness in the International Economy.* Washington, D.C.: Government Printing Office, 1984.

———. *64K Dynamic Random Access Memory Components From Japan Determination of the Commission in Investigation no. 731-TA-270 Under the Tarrif Act of 1930.* USITC Publication no. 1735. Washington, D.C.: International Trade Commission, January 1985.

———. *Dynamic Random Access Memory Semiconductors 256 Kilobits and Above From Japan: Determination of the Commission in Investigation no. 731-TA-300 Under the Tarriff Act of 1930.* USITC Publication no. 1803. Washington, D.C.: International Trade Commission, January 1986.

———. Global Competitiveness: Optical Fibers, Technology and Equipment. USITC Publication no. 2054. Washington, D.C.: International Trade Commission, January 1988.

———. *U.S. Trade Performance in 1987.* Washington, D.C.: Government Printing Office, 1988.

———. *Joint Report of the U.S.-Japan Working Group on the Structural Impediments Initiative.* Washington, D.C.: Government Printing Office, 1990.

———. *Global Competitiveness of U.S. Advanced Technology Manufacturing Industries: Semiconductor Manufacturing and Testing Equipment.* USITC Publication no. 2434. Washington, D.C.: International Trade Commission, 1991.

U.S. National Advisory Committee on Semiconductors. *A Strategic Industry at Risk.* Washington, D.C.: Government Printing Office, November 1989.

———. *Preserving the Vital Base: America's Semiconductor Materials and Equipment Industry.* Washington, D.C.: Government Printing Office, July 1990.

———. *Capital Investment in Semiconductors: The Lifeblood of the U.S. Semiconductor Industry.* Washington, D.C.: Government Printing Office, September 1990.

———. *Attaining Preeminence in Semiconductors.* Washington, D.C.: Government Printing Office, Februrary 1992.

———. *A National Strategy for Semiconductors: An Agenda for the President, the Congress, and the Industry.* Washington, D.C.: Government Printing Office, February 1992.

U.S. National Research Council. *Race for the New Frontier: International Competition in Advanced Technology.* New York: Simon & Schuster, 1984.

———. *U.S.-Japan Strategic Alliances in the Semiconductor Industry: Technology Transfer, Competition, and Public Policy.* Washington, D.C.: National Academy Press, 1992.

U.S. National Science Board. *Science and Engineering Indicators.* Washington, D.C.: National Science Board,1992.

U.S. National Science Foundation. *Technological Innovation and Federal Government Policy: Research and Analysis of the Office of National R&D Assessment.* Washington, D.C.: National Science Foundation, 1976.

———. *Research and Development in Industry.* Surveys of Science Resources Series.

Washington, D.C.: National Science Foundation, 1986.

U.S. Office of Technology Assessment. *International Competitiveness in Electronics*. Washington, D.C.: Government Printing Office, 1983.

———. *The Defense Technology Base: A Special Report*. Washington, D.C.: Government Printing Office, March 1988.

———. *Competing Economies: America, Europe, and the Pacific Rim*. Washington, D.C.: Government Printing Office, 1991.

U.S. President. *Economic Report of the President Transmitted to Congress: 1980*. Washington, D.C.: Government Printing Office, 1981.

———. *Economic Report of the President Transmitted to Congress: 1990*. Washington, D.C.: Government Printing Office, 1992.

U.S. Senate Committee on Finance. *Promotion of High Growth Industries and U.S. Competitiveness*. Washington, D.C.: Government Printing Office, 1983.

U.S. Technology Administration, U.S. Department of Commerce. *Emerging Technologies: A Survey of Technical and Economic Opportunities*. Washington, D.C.: Department of Commerce, Spring 1990.

Utterback, James M, and Albert E. Murray. *The Influence of Defense Procurement and Sponsorship of Research and Development of the Civilian Electronics Industry*. Cambridge: MIT Center for Policy Alternatives, 1977.

Vernon, Raymond, ed. *The Technology Factor in International Trade*. New York: Columbia University Press, 1970.

———. "Enterprise and Government in Western Europe." In *Big Business and the State*, Raymond Vernon, ed. Cambridge: Harvard University Press, 1981.

———. "A Strategy for International Trade." *Issues in Science and Technology* (Winter 1988): 73-79.

VLSI Research Inc. "The Top Ten Semiconductor Equipment Suppliers for 1991." *Electronic Business*. December 1991: 27.

Vogel, Ezra. *Japan as Number One: Lessons for America*. Cambridge: Harvard University Press, 1979.

Wald, Matthew. "Government Dream Car." *New York Times*. 1 October 1993: A1. *Washington Post*. 4 October 1987: H1.

Watson, William G. *A Primer on the Economics of Industrial Policy*. Ottawa: Ontario Economic Council, 1983.

Webbink, Douglas W. *The Semiconductor Industry: A Survey of Structure, Conduct, and Performance*. Staff report to the Federal Trade Commission. Washington, D.C.: Federal Trade Commission Bureau of Economics, 1977.

"What's Happening to British Chips." *The Economist*. 12 December 1987: 76.

White, Robert M. Personal interview with the Under Secretary of Technology. 12 June 1992.

———. "Technology and the Bush Administration: Moving Beyond the Industrial Policy Debate." Speech to the National Press Club. Washington, D.C. 29 September 1992.

Widenbaum, Murray, and Michael Athey. "What is the Rust Belt's Problem." In *The Industrial Policy Debate*, Chalmers Johnson, ed. San Francisco: Institute for Contemporary Studies, 1984.

Williams, Robert, and George Lodge. "SEMATECH." *Harvard Business School Case Study 389-057*. Cambridge: Harvard Business School, 1989.

Wolff, Alan W. "International Competitiveness of American Industry: The Role of U.S.

Trade Policy." In *U.S. Competitiveness in the World Economy*, Bruce Scott and George C. Lodge, eds. Cambridge: Harvard Business School Press, 1985.

World Economic Forum. *The World Competitiveness Report: 1990.* Geneva: EMF Foundation, 1991.

World Trade Data Base. "Technological and Trade Competition in High Technology Products." *BRIE Working Papers*. Berkeley: University of California Berkeley, 1991.

Yochelson, John, ed. *Keeping Pace: U.S. Policies and Global Economic Change*. Cambridge: Ballinger Publishing Company, 1988.

Yoshitomi, Masaru. "Keiretsu: An Insider's Guide to Japan's Conglomerates." *International Economic Insights* (September/October 1990): 50-58.

Young, John A. and others. *Global Competition: The New Reality*. The Report of the President's Commission on Industrial Competitiveness, Vol. II. Washington, D.C.: Government Printing Office, 1985.

Zysman, John. "Between the Market and the State: Dilemmas of French Policy for the Electronics Industry." *Research Policy* 7 (1978): 34-39.

———. *American Industry in International Competition: Government Policies and Corporate Strategies*. Ithaca: Cornell University Press, 1983.

———. *Governments, Markets, and Growth: Financial Systems and the Politics of Industrial Change*. Ithaca: Cornell University Press, 1983.

Index

About the Author

ERIC MARSHALL GREEN is an economist with the Federal Reserve Bank of New York and has worked at the U.S. Department of Treasury. He holds degrees from Holy Cross College, the University of London, and a doctorate from the Fletcher School at Tufts University.

ISBN 0-275-95253-3
90000>
EAN
9 780275 952532
HARDCOVER BAR CODE